885

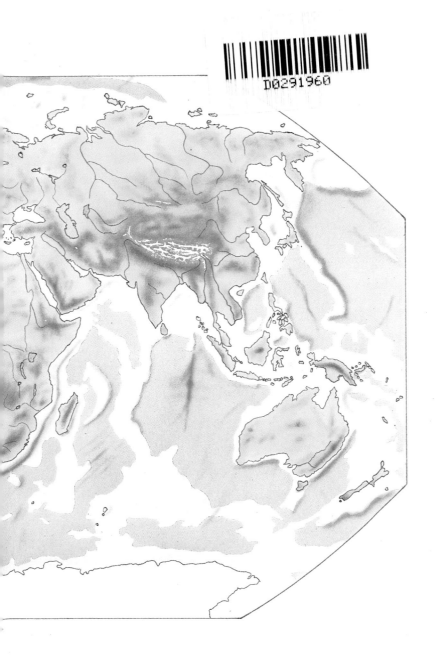

JOHN KINGSTON

LONGMAN ILLUSTRATED DICTIONARY OF GEOGRAPHY

the study of the Earth,
its landforms and peoples

LONGMAN   YORK PRESS

The author would like to acknowledge with thanks the contribution of
Dr R. J. Towse.

YORK PRESS
Immeuble Esseily, Place Riad Solh, Beirut.

LONGMAN GROUP UK LIMITED
Burnt Mill, Harlow, Essex.

First published 1988

ISBN 0 582 02163 4

Illustrations by Industrial Art Studio
Phototypeset in Britain by Input Typesetting Ltd, London, England.
Printed and bound in Lebanon by Typopress, Beirut.

Contents

How to use the dictionary

This dictionary contains about 1800 words used in geography. These are arranged in groups under the main headings listed on pp. 3–4. The entries are grouped according to the meaning of the words to help the reader to obtain a broad understanding of the subject.

At the top of each page the subject is shown in bold type and the part of the subject in lighter type. For example, on pp. 16 and 17

16 · **LITHOSPHERE**/STRUCTURAL GEOLOGY

LITHOSPHERE/PALAEONTOLOGY · **17**

In the definitions the words used have been limited so far as possible to about 1500 words in common use. These words are those listed in the 'defining vocabulary' in the *New Method English Dictionary* (fifth edition) by M. West and J. G. Endicott (Longman 1976). Words closely related to these words are also used: for example, *characteristics*, defined under *character* in West's *Dictionary*.

In addition to the entries in the text, the dictionary has several useful appendixes which are detailed in the Contents list and are to be found at the back of the dictionary.

---

1. To find the meaning of a word

Look for the word in the alphabetical index at the end of the book, then turn to the page number listed.

In the index you may find works with a number at the end. These only occur where the same word appears more than once in the dictionary in different contexts. For example, **family**

family[1] is a taxon;

family[2] is a group of closely related people.

The description of the word may contain some words with arrows in brackets (parentheses) after them. This shows that the words with arrows are defined near by.

(↑) means that the related word appears above or on the facing page;

(↓) means that the related word appears below or on the facing page.

A word with a page number in brackets after it is defined elsewhere in the dictionary on the page indicated. Looking up the words referred to may help in understanding the meaning of the word that is being defined.

In some cases more than one meaning is given for the same word. Where this is so, the first definition given is the more (or most) common usage of the word. The explanation of each word usually depends on knowing the meaning of a word or words above it. For example, on p. 33 the meaning of *ground moraine*, *lateral moraine*, *medial moraine*, and the words that follow depends on the meaning of the word *moraine*, which appears above them. Once the earlier words have been read those that follow become easier to understand. The illustrations have been designed to help the reader understand the definitions but the definitions are not dependent on the illustrations.

2. To find related words

Look in the index for the word you are starting from and turn to the page number shown. Because this dictionary is arranged by ideas, related words will be found in a set on that page or one nearby. The illustrations will also help to show how words relate to one another.

For example, words relating to Pollution are on pp. 197–202. On p. 197 *pollution* is followed by words used to describe the different problems of pollution and illustrations showing some of these problems; p. 198 explains and illustrates ecological disruption with entries on *eutrophication* and *ecological disruption*; p. 199 explains pollution control and pp. 200–202 gather together the remaining words on pollution with entries on many kinds of pollutants and their effects.

3. As an aid to studying or revising

The dictionary can be used for studying or revising a topic, or more simply to refresh your memory. For example, to revise your knowledge of demography, you would look up *demography* in the alphabetical index. Turning to the page indicated, p. 146, you would find *demography, population, population density, population distribution*, and so on; on p. 147 you would find *economic characteristics of population, population dynamics*, and so on; on p. 148 you would find *Lorenz curve, population spacing measure*, etc.

In this way, by starting with one word in a topic you can revise all the words that are important to this topic.

4. To find a word to fit a required meaning

It is almost impossible to find a word to fit a meaning in most dictionaries, but it is easy with this book. For example, if you had forgotten the word for the end point of a succession, all you would have to do would be to look up *succession* in the alphabetical index and turn to the page indicated, p. 76. On reading the definition you would be referred to *climax* on the previous page and there you would find the word.

5. Abbreviations used in the definitions

abbr	abbreviation	p.	page
adj	adjective	pl	plural
e.g.	*exempli gratia* (for example)	pp.	pages
etc	*et cetera* (and so on)	sing.	singular
i.e.	*id est* (that is to say)	v	verb
n	noun	=	the same as

THE
DICTIONARY

geography (*n*) the study and explanation of the forms, both man-made and natural, that make up the surface of the Earth. Special attention is given to the way that these forms are arranged in space. **geographic** (*adj*).

physical geography the study of the arrangement and causes of natural things such as soil (p. 88), vegetation (p. 86), climate (p. 71) and relief (p. 23) at and near the Earth's land surface.

lithosphere (*n*) the part of the Earth above the inner mantle (↓). It includes the solid crust (↓) and the outer mantle. The lithosphere is up to 100km thick. **lithospheric** (*adj*).

crust (*n*) the outer layer of the Earth, the lower boundary (p. 173) of which is marked by the Mohorovicic discontinuity (↓). It is between 20 and 40 km thick under the continents and 5 km under the oceans. **crustal** (*adj*).

mantle (*n*) the layer of the Earth between the crust (↑) and the core (↓). It acts like a solid and is about 2870 km thick. The mantle lies between the Mohorovicic (↓) and the Gutenberg (↓) discontinuities. It consists of minerals such as olivene and pyroxene.

asthenosphere (*n*) a zone (p. 241) of the mantle (↑) that lies directly beneath the lithosphere (↑). It is between 100 and 240 km below the Earth's surface.

core[1] (*n*) the dense nickel-iron centre of the Earth. The outer margin is marked by the Gutenberg discontinuity (↓). A solid inner core with a radius (distance from centre to edge) of 1400 km is recognized within the core itself.

section through the Earth

crust

upper mantle

Mohorovicic discontinuity

lower mantle

Gutenberg discontinuity

outer core

inner core

crust

a section through the Earth's crust

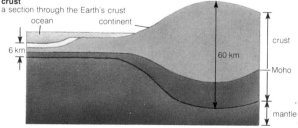

ocean   continent

6 km

60 km

crust

Moho

mantle

discontinuity (*n*) a boundary (p. 173) between two Earth layers across which the physical properties change markedly. If the velocity of earthquake (p. 208) waves changes, this is called a seismic (p. 208) discontinuity.

Gutenberg discontinuity one of the two major discontinuities in the Earth's structure. The Gutenberg discontinuity lies 2900 km down.

Mohorovicic discontinuity a seismic (p. 208) discontinuity (↑) that separates the crust (↑) above from the mantle (↑) below. It is found at a depth of 20–40 km below the surface of the continents and 5–10 km below the ocean floor (p. 101). The boundary (p. 173) is marked by a change in the velocity of earthquake (p. 208) waves. Also known as **Moho** or **M discontinuity**.

sial (*n*) the part of the crust (↑) found on the continental masses. It is made up of rocks rich in *si*lica (↓) and *al*uminium, such as granites (p. 10). *See also* sima (↓). **sialic** (*adj*).

silicate SiO₄

sima (*n*) the part of the crust (↑) found under the oceans. It is made up of basaltic (p. 10) rocks rich in *si*lica (↓) and *ma*gnesium. *See also* sial (↑). **simatic** (*adj*).

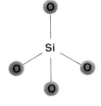

rock (*n*) a natural material formed of minerals. Rocks may be hard or soft, as in granites (p. 10) or muds, and can be divided into three groups: igneous (p. 10), metamorphic (p. 11) and sedimentary (p. 11). **rocky** (*adj*).

mineral (*n*) a natural substance with a definite chemical make-up. Most minerals have a characteristic crystal (↓) form.

quartz

crystal (*n*) a chemical substance with a well-defined geometrical form, with flat faces arranged at regular angles to one another.

silica (*n*) a very hard white substance; it is an oxide (p. 10) of silicon (Si), SiO_2, that is present in many minerals including quartz (↓). **siliceous** (*adj*).

silicate (*n*) a compound (p. 10) of silicon (Si), oxygen (O) and one or more metals; the most important rock forming minerals.

quartz (*n*) a hard, glass-like mineral; chemical composition, silica (↑).

oxide (*n*) a chemical compound (↓) of oxygen and another element (↓).

chemical compound a substance made up of two or more elements (↓).

element[1] (*n*) a substance that cannot be broken down by chemical means into simpler ones.

calcite (*n*) calcium carbonate, $CaCO_3$.

micas (*n.pl.*) layered or platy minerals that often have a book-like appearance. A good cleavage (p. 14) results in flakes of mica, muscovite being a colourless variety and biotite, brown or black.

clay minerals these are layered or fibrous (p. 13) in character. They are hydrous (water-containing) silicates (p. 9) of mainly aluminium and magnesium and are the common minerals of argillaceous (p. 12) or clayey (p. 12) rocks. They are formed by the weathering (p. 19) of silicate minerals.

igneous rocks those rocks which have solidified (become solid) from a magma (↓), either at depth or on the surface of the Earth. Igneous rocks are grouped according to the amount of silica (p. 9) present and on the size of the crystals (p. 9) or grains present.

magma (*n*) an under-surface liquid consisting of silicates (p. 9) and dissolved gases. As the hot magma cools, crystals (p. 9) grow and different rocks may be formed at different depths and temperatures. If the magma pours over the Earth's surface, lavas (↓) are formed.

granite (*n*) a coarse-grained igneous rock (p. 9) containing quartz (p. 9) silicates and micas (↑). Microgranites are the medium-grained equivalents of granites. **granitic** (*adj*).

basalt (*n*) a fine-grained volcanic rock (p. 9). **basaltic** (*adj*).

trachyte (*n*) a fine-grained intermediate igneous rock (p. 9) often found at plate margins (p. 15).

volcanic rocks rocks poured out over the Earth's surface from a chamber or space containing magma (↑); e.g. lava (↓).

lava (*n*) a molten liquid (i.e. magma (↑)) poured out over the Earth's surface. The rock formed when the liquid cools is also known as lava.

magma ascent of magma

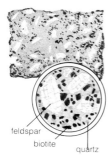
granite
coarse grained igneous rock

feldspar
biotite
quartz

lava

gneiss

marble

sandstone

metamorphic rocks rocks formed by changes in existing rocks within the Earth's crust (p. 9). These changes (metamorphism) may be a result of heat or pressure. **metamorphose** (*v*).

gneiss (*n*) a coarse-grained metamorphic rock (↑) with a foliation (p. 13) consisting of alternating layers of light and dark minerals.

schist (*n*) a medium- to coarse-grained metamorphic rock (↑) with a well-marked foliation (p. 13) due to the lines of layered platy minerals.

quartzite (*n*) a metamorphosed (↑) quartz-rich sandstone.

marble (*n*) a metamorphosed (↑) limestone (p. 12).

slate (*n*) a fine-grained argillaceous rock (p. 12) with a good cleavage (p. 14). **slaty** (*adj*).

sedimentary rocks rocks formed as the product of erosion (p. 20) or as a chemical precipitate (p. 14).

sediment (*n*) a material deposited by water, ice or wind, or possibly as a chemical precipitate (p. 14). **sedimentary** (*adj*).

arenaceous rocks sediments (↑) consisting of pieces from 1/16 mm to 2 mm in size. Most are of sandstone (↓) grade and many are made up of quartz (p. 9) grains (p. 13).

sandstone (*n*) an arenaceous (↑) sediment (↑) or clastic (p. 13) rock consisting of pieces 1/16 mm to 2 mm in size. They vary in mineral content and the grains (p. 13) may be held firm by a cement (↓).

cement (*n*) the material between the particles of a sedimentary rock (↑). It is this material that holds the particles together.

sand (*n*) a clastic (p. 13) deposit (p. 13) made up of pieces. 1/16 mm to 2 mm in size and not held firm by a cement (↑).

grit (*n*) medium- to coarse-grained arenaceous rock (↑).

rudaceous rocks sedimentary rocks (↑) with an average grain size greater than 2 mm, e.g. breccias (p. 12) and conglomerates (p. 12).

gravel (*n*) a rudaceous (↑) sediment (↑) in which the grain size varies between 2 and 4 mm. It is deposited mainly in river beds.

conglomerate (*n*) coarse- to very coarse-grained rudaceous rock (p. 11) consisting of rounded or subrounded (↓) clasts (↓). The clasts may be of any rock and may range in size from a few millimetres to over a metre.

subrounded (*adj*) of a grain which has had its corners and edges slightly rounded.

breccia (*n*) a coarse grained angular sedimentary rock (p. 11) with angular pieces.

argillaceous rocks sedimentary rocks (p. 11) with particles less than 1/16 mm in size; clays (↓), silts (↓) and mudstones (↓). They are a group of fine-grained clastic (↓) rocks.

clay (*n*) a fine-grained argillaceous rock (↑) with particles less than 1/256 mm in size.

shale (*n*) an argillaceous rock (↑) with clay (↑) size particles but a well-marked bedding (p. 18).

mudstone (*n*) an argillaceous rock (↑) similar to a shale (↑) but less fissile (↓).

silt (*n*) an argillaceous rock (↑) with particles between 1/16 mm and 1/256 mm in size.

carbonate rocks sedimentary rocks (p. 11) with a high calcium carbonate ($CaCO_3$) content. They may be deposited in sea water or fresh water as the result of the concentration (p. 106) of organic remains, the precipitation (p. 55) of calcium carbonate or the erosion (p. 20) of existing rock or organic materials.

limestones (*n.pl.*) a carbonate rock (↑) formed chiefly (>50%) of calcite (p. 10) or dolomite .(↓). Limestones can be divided into organic, clastic (↓) or precipitated (p. 14) groupings.

dolomite (*n*) a carbonate-magnesium mineral, $CaMg(CO_3)_2$. Also a carbonate rock (↑) with a high magnesium content formed as a replacement of limestones (↑).

carbonaceous rocks rocks which contain carbon. Carbonaceous deposits include peat (p. 92) and coal (↓).

coal (*n*) a carbonaceous (↑) deposit (↓) containing the remains of plants. Physical and chemical changes result in a hard, black substance that can be burned.

lithology (*n*) the general character of a rock as seen in an outcrop (p. 18) or hand specimen.

conglomerate

breccia

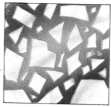

coal

texture
well-sorted, rounded and
closely packed sandstone
shows the texture of highly
mature sediments

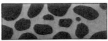

porous

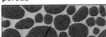

pore space in well-sorted
sandstone

porosity related to fracturing
of rock

pore spaces enlarged by
solution

grain

sedimentary rock	grain or particle size
clay	less than 0.005 mm
silt	0.005 – 0.05 mm
sand	0.05 – 2 mm
gravel	2 – 4 mm
pebble	4 – 65 mm
cobble	65 – 250 mm
boulder	more than 250 mm

clast

rock clasts

mineral clasts

organic clasts

schistocity

petrology (*n*) the study of the chemistry,
minerals and origins of rocks. **petrological** (*adj*).

texture (*n*) the relationships between minerals
or grains (↓) within a rock either in hand
specimen or thin section. **textural** (*adj*).

phenocryst (*n*) large crystals (p. 9) formed in
the first stages of cooling of an igneous rock
(p. 10).

matrix (*n*) fine-grained material that fills the
space between larger grains or crystals (p. 9)
in rocks.

fibrous (*adj*) of minerals that appear as threads
or fibres.

porous (adj) of a rock with holes or pore spaces
(↓). **porosity** (*n*).

pore space a space between the particles that
make up a rock (p. 9).

permeable (adj) of a rock through which liquids
and gases may pass. **permeability** (*n*).

impermeable (adj) not permeable.

fissile (adj) of rocks that split easily along closely
spaced parallel lines.

clastic (*adj*) of sedimentary rocks (p. 11) made
up of rock or mineral pieces. *See* clast (↓).

grain (*n*) a mineral particle, rock or fossil (p. 17)
pieces found in a rock. The texture (↑) of a
rock may be described as fine, medium or
coarse grained. **grained** (*adj*).

sorting (*n*) the degree to which the particles that
make up a rock (p. 11) or other rock material
are judged to be alike, e.g. in size or shape.

clast (*n*) a piece of rock or mineral within an
arenaceous rock (p. 11).

sedimentation (*n*) the process of forming
sediments (p. 11).

deposition (*n*) the process by which a sediment
(p. 11) is laid down; sedimentation. **deposit** (*n, v*).

alluvium (*n*) material deposited (↑) by rivers.

foliation (*n*) a layered, planar (p. 18) feature in
metamorphic rocks (p. 11) that results from
the secondary growth of minerals.

schistosity (*n*) a kind of foliation (↑) that is due
to the parallel or almost parallel arrangement
of minerals (e.g. micas (p. 10)) in schists
(p. 11) and gneisses (p. 11).

cleavage (n) the condition when a rock tends to split easily along parallel planes or surfaces. A slaty cleavage is found in fine-grained rocks which have been under great pressure. *Fracture cleavage* may be found in rocks that are deformed (i.e. changed in shape). *Flow cleavage* may be the result of a recrystallization (i.e. the dissolving and regrowth of crystals (p. 9) of minerals) in a pre-existing rock.

precipitate (n) a solid that appears from solution as a result of a chemical reaction (↓).

chemical reaction this takes place when two substances have an effect on each other and a new substance is produced.

solution (n) a liquid in which substances (solids) are dissolved (broken down into molecules or ions).

structural geology the study of the shapes and positions of rock masses. Geology is the study of the Earth.

structure (n) the way in which the parts of a thing are arranged, for example, a system or the shape and positions of rock masses; the form of a geometric shape, a fold (↓) or a fracture (break). **structural** (adj).

anticline (n) a fold (↓) shaped like an arch; a fold in which the oldest rocks are in the centre. **anticlinal** (adj).

syncline (n) a fold (↓) shaped like the letter U; a fold with the youngest rocks in the centre. **synclinal** (adj).

fault (n) a break in rocks along which movement has taken place.

joint (n) a break in rocks along which no movement has usually taken place.

fold (n) a bend in rocks e.g. an anticline (↑) or syncline (↑). **fold** (v), **folded** (adj).

plate (n) a large block of crustal (p. 8) material that 'floats' over the asthenosphere (p. 8). The plate may be made up of continental or oceanic materials. Africa, North and South America, Antarctica, Eurasia, India and the Pacific are the seven main plates. There are also a number of small ones.

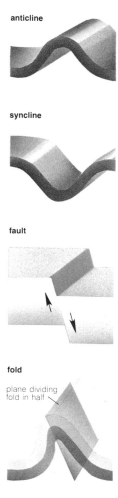

anticline

syncline

fault

fold

plane dividing fold in half

plate tectonics a geological theory (p. 222) put forward during the 1960s. It is concerned with the movement of the relatively thin plates (↑) that make up the surface of the Earth. The plates reach down into the upper part of the mantle (p. 8). Each is edged by plate margins (↓). The theory includes the idea of sea-floor spreading in which magma (p. 10) rising along mid-oceanic ridges forms new crust (p. 8). Repeated rifting (↓) and the continued addition of magma results in the sideways movement of older material. The addition of new crustal materials at the mid-oceanic ridge is compensated for by the subduction (↓) of the crust at destructive plate margins (↓). As an oceanic plate expands or spreads, the neighbouring continental plates are moved across the Earth's surface.

plate margins the edges of a plate (↑). There are three kinds of margin. (1) *constructive*: where new crustal (p. 8) material is added to the plates on either side; e.g. the mid-Atlantic ridge. (2) *destructive*: where the ocean floor is destroyed through subduction (↓). Deep oceanic trenches (p. 50) often mark the presence of a destructive margin. (3) *passive* or *conservative*: where two plates slip past each other without the addition or destruction of crustal material, often marked by massive tear or transform faults (p. 16).

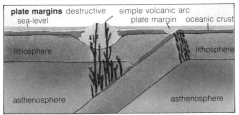

plate margins destructive — simple volcanic arc — plate margin — oceanic crust — sea-level — lithosphere — lithosphere — asthenosphere — asthenosphere

subduction (*n*) the process by which oceanic crust (p. 101) is destroyed within a destructive plate margin (↑). It takes place when one plate (↑) rides over another and forces crustal material down into the mantle (p. 8).

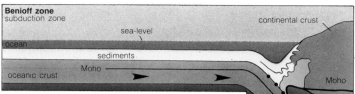

Benioff zone
subduction zone

sea-level

ocean

sediments

Moho

oceanic crust

continental crust

Moho

base of plate

Benioff zone

asthenosphere

Benioff zone a seismic (p. 208) zone (p. 241) that stretches downwards from an oceanic trench (p. 50). Named after H. Benioff, it marks the boundary (p. 173) between two plates (p. 14) and slopes at an angle of 45° from crust (p. 8) to asthenosphere (p. 8).

shield area a very large unit of the Earth's crust (p. 8) made up of Precambrian rocks. Shields form the central parts of continental plates (p. 14) and have remained unchanged by mountain-building episodes over a long stretch of geological time. The Canadian and the Baltic shield of Northern Europe are examples of such ancient stable blocks.

island arc a long chain of islands that are bordered on their oceanic side by a deep trench (p. 50). The chain is bow-shaped or curved and marked by volcanoes (p. 40). The Aleutians and Marianas are examples of island arcs in the Pacific.

transform fault a large fault (p. 14) between two plates (p. 14), at right angles to the mid-oceanic ridges (p. 101).

horst (*n*) a block of rocks between two faults (p. 14) which has moved upwards relative to the rocks around it.

graben (*n*) a block of rocks between two parallel faults (p. 14) which has moved downwards relative to the rocks around it.

rift (*n*) a long structural feature marked by two parallel faults (p. 14). A rift is both a structural and topographical feature, usually marked by a deep flat-bottomed valley.

Pangea (*n*) a former 'super-continent' consisting of all the Earth's land masses, which began to break up about 200 million years ago. *See also* Gondwanaland (↓) and Laurasia (↓).

shield area
map of Africa showing the shield areas

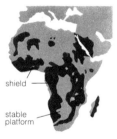

shield

stable platform

horst

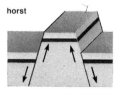

graben

Tethys

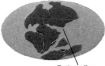

Tethys Sea

isostasy
two possible models

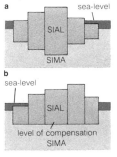

fossil

Gondwanaland (*n*) the name given to an ancient land mass, consisting of the southern continents, which resulted from the break-up of Pangea (↑). It is important biogeographically (p. 72) because many taxonomic (p. 77) groups are thought to have arisen there, later to spread by various dispersal routes (p. 84) or by further continental break-up.

Laurasia (*n*) the name given to an ancient land mass of the northern hemisphere (p. 238) which resulted from the break-up of Pangea (↑).

Tethys (*n*) an ocean that once stretched from the Caribbean, through the Mediterranean into the Indian Ocean. Tethys first opened in the late Palaeozoic and closed during the Tertiary.

nappe (*n*) a large structural feature. It is a large, horizontal fold that has been pushed tens of kilometres away from its original position.

isostasy (*n*) the theory (p. 222) that all the larger blocks of the Earth's crust (p. 8) are floating on a denser underlying material; with erosion (p. 20) and deposition (p. 13) the equilibrium (p. 220) between these blocks is upset and this is countered, according to the theory, by earth movements. Areas of erosion rise, areas of deposition sink. There are two models of isostasy. Airey assumes that all sialic (p. 9) rocks have the same density and the difference in elevation of the blocks is due to their differing thickness (**a**). Pratt assumes that the higher blocks have a lower density than the lower blocks and so there is a level – the level of compensation – to which all blocks sink (**b**). *See diagram.* **isostatic** (*adj*).

palaeontology (*n*) the study of fossils. **palaeontological** (*adj*).

fossil (*n*) the remains or traces of an animal or plant preserved in rock. *Body fossils* include the minerals, shells or skeletons (bony frames) of animals; *trace fossils* show how the animals lived, e.g. tracks and burrows (holes made by the animals).

chalk (*n*) a soft white rock formed of plant and animal remains.

palaeoecology (n) the study of fossil (p. 17)
    animals and plants in relation to the conditions
    under which they lived. This includes a study
    of the sediment (p. 11) that contains the
    fossils. **palaeoecological** (adj).
palaeobiogeography (n) the study of the way
    animals and plants were distributed over the
    Earth's surface in past times.
    palaeobiogeographical (adj).
outcrop (n) an area of rock on the surface of the
    Earth. An outcrop is usually material that can
    be seen but the name may be used where the
    rock is covered by drift (p. 33).
stratum (n) = a bed (↓). **strata** (pl).
bedding plane a surface between beds (↓)
    parallel to the original surface on which the
    sediment (p. 11) was deposited (p. 13).
plane (n) a flat surface, although surfaces such
    as bedding planes (↑) described by
    geologists may be rather uneven. **planar** (adj).
stratiform[1] (adj) of rocks their layered nature.
bed (n) a layer of sedimentary rock (p. 11) that is
    marked above and below by a break in
    deposition (p. 13). **bedding** (n).
dip (n) the angle a bedding plane (↑) makes with
    a horizontal plane. The measurement is taken
    at right angles to the strike (↓) of the rock.
strike (n) the direction of a horizontal line on a
    bedding plane (↑) taken at right angles to the
    dip (↑).

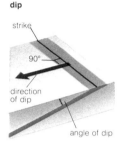

dip

strike

90°

direction
of dip

angle of dip

bedding plane

bed

bedding plane

geomorphology (*n*) the study of the Earth's relief (p. 23) (including that under the sea) and of the various ways in which it has been produced. **geomorphological** (*adj*).

geomorphological process the action of various forces that bring about change at or near the Earth's surface. The forces may be due to the work of rivers, glaciers (p. 28), wind, waves and weathering (↓).

weathering (*n*) the processes that cause rock to break up in place. These include the action of water containing weak acid on minerals and the break up that results when water freezes and expands in joint (p. 14) spaces.

physical weathering the natural break up of rock into smaller pieces without changing its nature. It includes freeze-thaw action (p. 36).

chemical weathering the break up of rock into smaller pieces by forces which change the nature and arrangement of the molecules.

biological weathering a group of weathering (↑) processes that are due to the action of plants and animals. These may produce weak acids that bring about chemical weathering (↑). Such acids are also produced when vegetation (p. 86) decays. Plants may bring it about when their roots grow in joints (p. 14) and force rocks apart.

exfoliation

exfoliation (*n*) a process by which layers may expand and lift away from a rock surface, as a result of weathering (↑). Also known as **onion skin' weathering**.

hydration (*n*) a kind of chemical weathering (↑) in which molecules of water stick electrically to mineral surfaces and the mineral increases in size and is weakened. **hydrate** (*v*).

taffoni (*n.pl.*) the small hollows in a rock surface that result from weathering (↑).

desert varnish a skin of material on many rocks in hot desert areas. It is made up of minerals that result from weathering (↑) and which are laid down at the surface by evaporation (p. 94).

regolith (*n*) the surface sheet of soil and broken rock that lies on sound rock below. It is mainly the result of weathering (↑).

duricrust (*n*) a hard layer of material that results when the minerals produced by weathering (p. 19) are laid down at some depth below the land surface. The layer may then appear at the surface because of erosion (↓).

sarsen (*n*) a boulder of sandstone (p. 11) held together by silica (p. 9) and formed as part of a duricrust (↑) on the Chalk (p. 17) of southeast England.

erosion (*n*) a general word for the various ways in which rock is broken and moved over a surface, e.g. by water, wind, ice and mass movement (↓). **erode** (*v*), **erosive** (*adj*).

soil erosion the loss of soil from a land surface by erosion (↑). It may be brought about by the growth of gullies (p. 28), by deflation (p. 38), or by the action of water when it flows as a sheet over the land. It is especially important in areas of low rainfall.

abrasion (*n*) the work done when particles of rock are moved over a land surface and so wear it down. The particles are usually carried by wind, water or ice. **abrade** (*v*).

saltation (*n*) a process by which rock particles are lifted by either water or wind, carried along and then laid down. The angle of lift is quite steep. **saltate** (*v*).

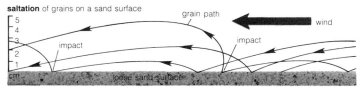

saltation of grains on a sand surface

mass movement the group of processes by which material moves in mass down hill slopes and sea cliffs. It includes landslides (↓), creep (↓) and the fall of rock.

creep (*n*) the slow movement of the upper part of the regolith (p. 19) down a hill slope. It is a result of gravity acting on small chance movements, e.g. those caused by frost heaving (p. 36). Soil creep is the most common form of this process.

landslides in coastal cliffs near Folkestone, Kent. The curved slip surfaces make this the 'rotational' type

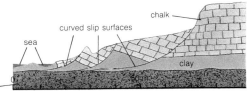

landslide (*n*) the sudden slip of part or all of a hill slope. It may be caused by undercutting at the base of the slope, by a rise in the water-table (p. 94) or by an increase in the weight of the slope material.

lahar (*n*) a flow of wet material down the side of a volcano's (p. 40) ash cone. It may happen after heavy rain.

scree

scree (*n*) a pile of broken angular rocks that rests against the base of a steep slope or cliff. It is often a result of freeze-thaw action (p. 36) on the slope above.

badlands (*n.pl.*) a landscape (p. 22) of many closely spaced gullies (p. 28) with steep sides that meet as sharp ridges (p. 43). There is often little plant cover and the rainfall is low.

cycle of erosion the changes that a landscape (p. 22) goes through from the time it was raised above sea-level until its destruction by erosion (↑). The old landscape is then raised to start the cycle again. The idea was put forward by W. M. Davis in 1899. The time needed for a cycle (65 million years) means that a change of climate (p. 71) or a period of earth movements is likely to take place before the cycle is complete.

cycle of erosion
steps in W. M. Davis's cycle of erosion

➡ = direction of main erosion

- - - starting
surface

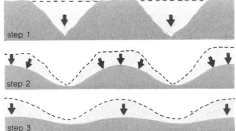

erosion surface a land surface made up of very
gentle slopes and which is the result of long
continued erosion (p. 20). Many different
processes may have taken place, e.g.
weathering (p. 19) and mass movement (p. 20).

plateau (*n*) a raised, nearly level part of the
Earth's land surface.

peneplain (*n*) the erosion surface (↑) that results
at the end of a cycle of erosion (p. 20). It is
produced by weathering (p. 19), mass
movement (p. 20) and river action.

part of a peneplain

monadnock

flood plain with
meandering
river

sea

etchplain (*n*) an erosion surface (↑) which
results when the regolith (p. 19) is moved
away by erosion (p. 20) and the rock floor
below then appears as the new land surface.

panplain (*n*) an erosion surface (↑) that results
from sideways erosion (p. 20) by rivers. It is
made up of many flood plains (p. 26) that are
joined together.

monadnock (*n*) a hill which stands well above
the general level of a peneplain (↑), often
because it is made of hard rock. Named after
Mount Monadnock, New Hampshire, USA.

base level that level below which erosion (p. 20)
ceases. For a large area the base level is sea-
level, while for a hill slope it is the river at its
foot.

rejuvenation (*n*) an increase in the speed of
erosion (p. 20) that happens when the sea-
level falls or when the land is raised.

consequent stream a river whose path has been
determined by the shape of a new land surface
rather than by rock structure.

landscape (*n*) the form of the land as seen from
a certain point, often including the effect of
plants and of man-made features.

topography (*n*) the surface forms of a region
(p. 241), including those that are man-made
those that are natural. **topographic** (*adj*).

rejuvenation
rejuvenation by a stream,
which has cut down from an
earlier, higher level (dotted
line)

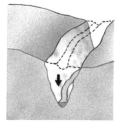

channel
relation between channel shape and sediment character for the Great Plains, USA

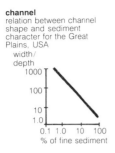

hydraulic radius

hydraulic
radius $\mathbf{R} = \dfrac{\mathbf{A} \text{ (area)}}{\mathbf{P} \text{ (wetted perimeter)}}$

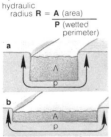

channel **a** (**R** = 0.56) is more effective than channel **b** (**R** = 0.4) in carrying water

pool and riffle

where pools and riffles are found in a meandering channel

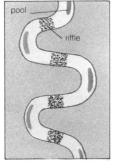

relief (n) the various natural shapes of the Earth's surface.

corrasion (n) the abrasion (p. 20) of a surface by pieces of rock that are drawn over it, especially by water. It occurs below the sea and in river channels (↓). It also takes place in rocky deserts (p. 87), where the work is done by sand-carrying wind. **corrade** (v).

attrition (n) the wearing down and rounding of the pieces of rock that are carried by a moving body, especially a river. In this case a result is that the bedload (↓) becomes finer downstream (towards the mouth).

suspension (n) the state of a piece of rock when it is both held up and moved along by flowing water such that it does not touch the sides of the channel (↓). **suspend** (v).

traction (n) the process by which relatively large pieces of rock are pulled along the floor of a river channel (↓) by the flowing water.

bedload (n) the sediment (p. 11) that is carried along the bed of a river.

channel (n) the bed and sides of a course, e.g. of a river. In this case, its shape may be related to the size of sediment (p. 11): large materials give rise to channels that are wide and not very deep, while narrower and deeper channels are found in finer sediments. *See also* hydraulic radius (↓).

wetted perimeter that part of a river channel (↑) that is in contact with the water. Its length varies with changes in discharge (p. 98).

hydraulic radius a measure of the effectiveness of the shape of a river channel (↑) for carrying water. It is worked out by dividing the area across a channel by the wetted perimeter (↑).

pool (n) a relatively deep area in a river channel (↑), found in the curve of a meander (p. 25) and straight channels, repeated at a distance of between 5 and 7 times channel width.

riffle (n) a shallow part of a river channel (↑) and which is found in both straight and meandering (p. 25) cases. It is usually made of sediments (p. 11) of relatively large size, and repeated at distances of 5 to 7 times channel width.

pothole (*n*) a small, deep, nearly circular hole found in the floor of a rocky river channel (p. 23). It is eroded (p. 20) by abrasion (p. 20) and the rounded pieces of rock ('tools') that did this are often found inside.

thalweg (*n*) (1) a line that joins the points of greatest water depth along a river channel (p. 23); (2) the line that joins the lowest points along a valley

pothole
pothole with sand and rock 'tools'

long profile long profiles of two English rivers

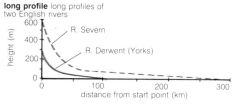

long profile the surface shape of a landform when measured in its longest direction. It often refers to a river, when it is obtained by noting the height of the water surface at increasing distances from the starting point.

knick point a change in the bed slope along the line of a river. It may be due to a change in the kind of rock, or to the down-cutting that happens when the sea-level falls.

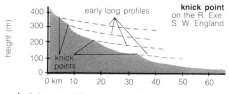

knick point
on the R. Exe. S. W. England

grade (*n*) a word first used by W. M. Davis in 1902 to describe a state of equilibrium (p. 220) between erosion (p. 20) and sedimentation (p. 13) in a river channel (p. 23). A result is that the form of the slope stays much the same for a long time. The long profile (↑) of a stream is said to be graded when it is smoothly curved. This is an old word which has come under attack in the last few years. **graded** (*adj*).

meander a meandering river: the Mississippi at Greenville, Mississippi, USA

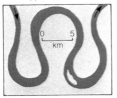

oxbow lake north of Llanrwst, N. Wales

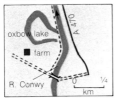

incised meander difference between ingrown (**a**) and entrenched (**b**) incised meanders

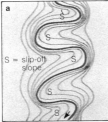

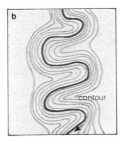

braided river the braided White River. It flows on outwash from the Emmons glacier, Mount Rainier, Washington state, USA

braided river a river consisting of many channels (p. 23). It flows down a relatively steep slope, has a discharge (p. 98) that shows large changes, and runs through sediments (p. 11) of differing sizes. It is often found on outwash plains (p. 35) where discharge varies and a range of sediment sizes occurs.

meander (*n*) a regular curve produced by a river and whose sinuosity (↓) is at least 1.5. It is formed most easily in fine flood plain (↓) sediment (p. 11). **meandering** (*adj*).

oxbow lake a lake in a curved length of channel (p. 23) which has been cut off from the river.

sinuosity
sinuosity of a meander is the down channel distance divided by the down valley distance.

In this case
$\frac{C_D}{V_D}$ = 1.57

down channel distance C_D

down valley distance V_D

sinuosity (*n*) the degree of curvature in plan of a river channel (p. 23). It is measured as the relation between the distance separating two points along the deepest part of the channel, and the straight-line distance between those points. A perfectly straight channel has a sinuosity of 1, while that of a meander (↑) is at least 1.5. **sinuous** (*adj*).

incised meander a meander (↑) that has cut well down into the rock beneath. Two varieties are found. The first is the *ingrown meander* which shows movement sideways and has a slip-off slope (p. 26), e.g. River Wye near Chepstow, Gloucestershire, England. The second is the *entrenched meander* which shows effectively no sideways movement, e.g. River Wear at Durham, northern England.

slip-off slope a slope on the side of a valley. It has been left behind by a river as it moved sideways while cutting down from a higher level. It is characteristic of an incised meander (p. 25). Good examples are found along the River Rhine, West Germany.

slip-off slope of the R. Rhine in a large meander. The river moved sideways in the direction of the arrows

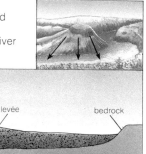

flood plain the part of a valley floor which a river may flood (p. 99) from time to time.

flood plain and levée
river flood plain and levées

flood plain · river · levée · bedrock

levée (*n*) a ridge (p. 43) of sediment (p. 11) along the bank of a river. It consists of material laid down when the river flows over its flood plain (↑)

river terrace
river terraces cut in drift, New England, USA

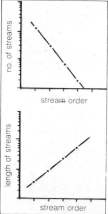

river network analysis
some results of a river network analysis

no. of streams / stream order

length of streams / stream order

river terrace a part of a former flood plain (↑) that now stands above the river channel (p. 23). It has been raised because the river has cut down its channel, perhaps because of a fall in sea-level, because flow conditions have changed, or because the land has risen.

drainage basin the area that supplies water to a river network (↓).

river network the way that all the streams of a drainage basin (↑) are arranged.

river network analysis the study of how stream order (↓) is related to a number of stream properties such as length and number.

stream order two ways of working out stream order

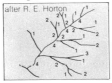

after R. E. Horton

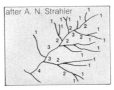

after A. N. Strahler

tributary R. Wey is a tributary of the R. Thames

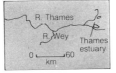

bifurcation ratio bifurcation ratio of this river is 17

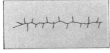

superimposed drainage how it may come about

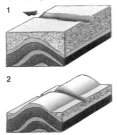

stream order a measure of the ranking order of a length of stream between two tributaries (↓) within the total river network (↑).

tributary (*n*) a stream that joins a river and which is usually smaller than the river it enters. It may be part of a river network (↑).

drainage density a measure of the average distance between the streams that flow across a land surface. It is calculated by adding together the lengths of all the streams and then dividing by the land surface area.

bifurcation ratio the number of streams of a certain order (*see* stream order (↑)) divided by the number of streams of the next higher order. If the ratio is a low number, the drainage basin (↑) may often flood (p. 99).

river capture in Northumberland, England. The broken lines show how streams were joined before capture by the North Tyne

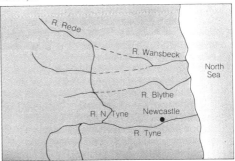

river capture the act by which one river takes over the discharge (p. 98) of another. It often happens when a river erodes up-valley, perhaps along a line of weak rock, and then cuts into a nearby stream that flows across its path.

superimposed drainage a stream (or streams) laid down on a cover of rocks quite different from and well above those now found at the surface of a certain area. Over time the stream eroded (p. 20) the cover and cut down to the different structures below, through which it now flows, e.g. many of the streams of Wales.

alluvial fan a sheet of sediment (p. 11) that spreads out from a point and which is laid down by a small river at the angle between a steep and gentle slope.

gully (*n*) a small steep-sided stream channel (p. 23) whose flow is not continuous and whose head erodes (p. 20) actively. A gully may grow quickly, especially when the soil or plant cover has been damaged.

arroyo (*n*) a Spanish word for a stream channel (p. 23) formed under a dry climate (p. 71). It is wide, has a flat floor and steep sides, and a steep long profile (p. 24). It is usually dry, and is floored with sediments (p. 11) of large size. The similar landform in Africa and the Middle East is known as a wadi.

delta (*n*) an area at or just beyond the mouth of a river where the build up of sediment (p. 11) is faster than its loss by erosion (p. 20). The best examples occur in seas whose level changes little (e.g. the Nile delta in the eastern Mediterranean) and in lakes (e.g. the Rhone delta in Lake Geneva). **deltaic** (*adj*).

alluvial fan

fan

delta Nile Delta, Egypt

old channels

beaches and spits

salt marsh

river sediment

high land

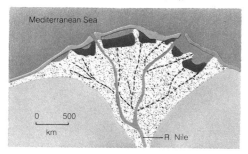

Mediterranean Sea

0   500
km
R. Nile

glacier (*n*) a body of ice usually containing pieces of rock and which moves under its own weight. It may be able to erode (p. 20), and to carry and lay down sediment (p. 11). Glaciers may be classified by their shape, e.g. ice sheet (↓), ice cap (↓), piedmont glacier (↓), valley glacier (↓), cirque glacier (↓), or by how cold they are, e.g. polar glacier (p. 30), temperate glacier (p. 30). **glacial** (*adj*).

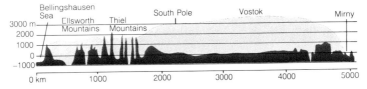

ice sheet Antarctic ice sheet

ice sheet a very large glacier (↑) that covers a relatively low-lying area, e.g. the Antarctic ice sheet that contains 85% of the world's ice.

ice cap a large glacier up to about 50,000 km² and which may be found in mountain areas (e.g. Vatnajökull ice cap in Iceland) or in lowland areas (e.g. the Barnes ice cap of Baffin Island, Canada).

valley glacier a tongue of ice held within the sides of a valley. It may be fed from an ice cap (↑) or from cirque glaciers (↓).

piedmont glacier a glacier (↑) that forms when one or more valley glaciers (↑) leave a mountain area and spread out over a nearby lowland, e.g. the Malaspina Glacier, Alaska, USA.

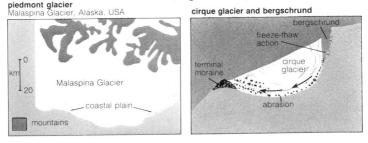

piedmont glacier
Malaspina Glacier, Alaska, USA

cirque glacier and bergschrund

cirque glacier a glacier (↑) that lies in a rocky (p. 9) hollow shaped like an armchair and usually less than 1 km across. It moves mainly by sliding, and this action forms the rocky hollow or cirque (p. 31) by abrasion (p. 20).

bergschrund (n) a long narrow opening found at the upper end of a cirque glacier (↑). It opens in summer when the head of the glacier pulls away from the steep wall of the cirque (p. 31) or from ice that sticks to it.

polar glacier a glacier (p. 28) whose
temperature is well below melting point. It is
frozen to its bed, moves slowly, and so may
not bring about much erosion (p. 20). Part of
the Greenland ice sheet (p. 29) has these
characteristics. Also known as **cold glacier**.

temperate glacier a glacier (p. 28) whose
temperature is close to the pressure melting
point. Meltwater is present throughout much
of the ice, and the glacier slides easily and
rapidly on its bed, often causing much erosion
(p. 20) by abrasion (p. 20). Many of the
glaciers of the Alps are temperate.

ablation (*n*) the ways by which snow and ice are
lost from a glacier (p. 28). It includes melting,
evaporation (p. 94) and the breaking away of
large masses of ice where the glacier enters
the sea ('calving'). **ablate** (*v*).

glacial period a part of the Quaternary when
large ice sheets (p. 29) covered much of the
middle latitudes (p. 238). The latest glacial
period in Britain lasted for about 100,000
years, but it was broken by interstadials (↓).

interglacial period the time between two glacial
periods (↑) when the climate (p. 71) became
warm enough for deciduous (p. 79) trees to
grow widely.

interstadial period a warm time during a glacial
period (↑) but one which was either too short
or too cold for deciduous (p. 79) trees to grow
widely.

glacial erosion the erosional (p. 20) work done
by a glacier (p. 28). It may include abrasion
(p. 20), plucking (↓), freeze-thaw action
(p. 36) and erosion by water due to melting.

plucking (*n*) the erosion (p. 20) carried out by a
glacier when it freezes to rock which is then
pulled away as the glacier moves on. It is
important when the rock has many joints (p. 14).

roche moutonnée

plucking

direction of ice flow

abrasion

striation
striations on a piece of rock

meltwater channel
meltwater channel next to a
glacier

crag and tail   crag
Edinburgh,   (Castle
Scotland   Rock)

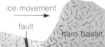

ice movement

fault

tail (High Street)

fault   hard basalt

soft sedimentary rocks

glaciated upland   pyramidal
some landforms   peak
arêtes   cirque

hanging valley

alluvial fans

striation (*n*) a long narrow cut in a rock surface.
It happens when a piece of rock, held in the
lower part of a glacier (p. 28), is pulled over a
rock surface.

meltwater channel a channel (p. 23) cut by
water escaping from a melting glacier (p. 28).
The water may flow in various places (e.g.
beside, under, or on top of, the glacier) and
gives rise to differences in the type of channel.

nunatak (*n*) a rocky hill that rises above the ice
sheet that surrounds it and which is eroded
(p. 20) by freeze-thaw action (p. 36).

roche moutonnée (*French*) a small rocky landform
produced by erosion (p. 20) under a glacier
(p. 28). The slope that faces the glacier is
quite smooth and gentle because of abrasion
(p. 20) while the other slope is steep and
rough because of freeze-thaw action (p. 36).

crag and tail a landform made up of a small
rocky hill (the crag) which is a result of erosion
(p. 20) by a glacier (p. 28), and a low sloping
ridge (p. 43) (the tail) made up of drift (p. 33)
laid down by the same glacier when it melted.
For example in Scotland, Edinburgh Castle
(on the crag) and the Royal Mile (on the tail).

cirque (*n*) a hollow, with a very steep slope at its
back, found in a mountainous area. Its slope
is due to the action of a glacier (p. 28), and
where this has melted a lake may be found,
often held back by a moraine (p. 33). Also
cwm (Wales), **corrie** (Scotland), **botn** (Norway).

pyramidal peak a mountain shaped like a horn
(e.g. the Matterhorn in Switzerland) which
results when the back walls of several cirques
(↑) come close together.

arête (*n*) a sharp, knife-edged ridge (p. 43) with
steep sides that is usually caused by both
glacial erosion (p. 30) and freeze-thaw action
(p. 36). It is often found in mountain areas
that were once covered by ice.

hanging valley a side valley whose lower end drops steeply down to the main valley. It often results from glacial erosion (p. 30) in the main valley being greater than the erosion (p. 20) by ice in the side valley. When the glaciers melt a difference in height is seen, and this may be marked by a waterfall.

glacial stairway
long profile of a glacial valley in West USA, showing rock steps

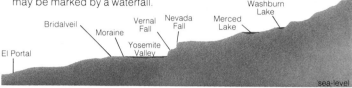

Mt. Lyell
3990 m
(13.090')

Washburn Lake

Nevada Fall

Vernal Fall

Merced Lake

Bridalveil

Moraine

Yosemite Valley

El Portal

sea-level

glacial stairway a valley floor made up of steep, rocky parts (steps), and those much less steep. This change in steepness results from differences in the amount of glacial erosion (p. 30) that once moved down the valley.

knock and lochan the name for a landscape (p. 22) made up of small rocky hills (knocks) and small lakes (lochans). It is a result of erosion (p. 20) by an ice sheet (p. 29), controlled by changes in rock strength.

glacial breach a side valley eroded (p. 20) by a glacier (p. 28) through a low point in a ridge (p. 43). This often happens when a glacier builds up faster than it can escape down valley and then flows over a nearby ridge.

glacial diffluence glacial breaching (↑) of watersheds (p. 93) over a large area, when many valleys become filled with ice. It may lead to big changes in the arrangement of the river system of the area.

glacial breach
its formation

valley glacier

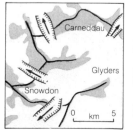

Carneddau

Glyders

Snowdon

0   km   5

glacial diffluence
glacial breaches in the Snowdon area, North Wales

land over 575 m

glacial breach with ice direction

watershed before arrival of glaciers

drift (*n*) the sediment (p. 11) laid down by the various activities of a glacier (p. 28). There are many kinds, for example those that result from simple melting under the glacier or from the laying down of material in a lake held in by ice.

till (*n*) the group of sediments (p. 11) laid down by the action of a glacier (p. 28) alone; water plays no part.

boulder clay till (↑). The name shows that many tills are made up of large blocks of rock (boulders) set in much finer material (e.g. clay (p. 12)). It lacks sorting (p. 13) and bedding (*see* bed (p. 18)), although single boulders may point in the direction of ice movement.

erratic
dark erratic of gritstone resting on the country rock of light-coloured limestone in the North Pennines, England

erratic (*n*) a rock carried and then dropped by a glacier (p. 28) and whose lithology (p. 12) is not the same as that of the country rock below. Its lithology may be used to work out the path of the glacier that laid the erratic down.

moraine (*n*) (1) a ridge (p. 43) or sheet of till (↑) laid down next to a glacier (p. 28). Several kinds are found, depending on their relation with the glacier. (2) the sediment (p. 11) laid down directly by a glacier.

ground moraine a sheet of sediment (p. 11) laid down beneath a glacier (p. 28).

lateral moraine a ridge (p. 43) of till (↑) laid down along the sides of a valley glacier (p. 29). Its material is supplied partly by the glacier and partly from the valley slopes above. It may show signs of water activity (*see* meltwater channel (p. 31)).

medial moraine a ridge (p. 43) of till (↑) laid down where two valley glaciers (p. 29) meet and their separate lateral moraines (↑) join.

moraine some types in a valley glacier

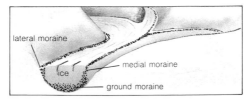

lateral moraine

medial moraine

ice

ground moraine

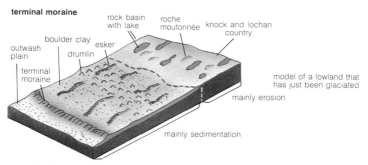

terminal moraine

rock basin with lake

roche moutonnée

knock and lochan country

boulder clay   esker

outwash plain

drumlin

terminal moraine

model of a lowland that has just been glaciated

mainly erosion

mainly sedimentation

terminal moraine a ridge (p. 43) of till (p. 33) laid down at the end of a glacier (p. 28).

ablation moraine a thin sheet of till (p. 33) laid down as a glacier ablates (p. 30). It usually lies on ground moraine (p. 33).

drumlin (n) a small hill with smooth sides, often made of drift (p. 33). Its shape, with one end sharp and the other less so, is a result of the way that a glacier (p. 28) flowed round and over it. It may be found on its own but is more often in a group. This produces the so-called 'basket of eggs' landscape (p. 22).

drumlin field of drumlins, showing the 'basket of eggs' appearance

esker (n) a long, low twisting ridge (p. 43) with steep sides and whose height changes smoothly. It is made of sediment (p. 11) laid down in a stream which once flowed inside or under a melting ice sheet (p. 29).

kame (n) a small hill made of drift (p. 33) laid down by water. Its sides were often supported by ice; when this melted they slipped down.

kame terrace a flat-topped ridge (p. 43) of drift (p. 33) laid down along the side of a valley by a stream flowing along the edge of a glacier (p. 28).

esker

kame terrace steps in formation

1

water-lain sediments at the sides of a glacier

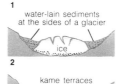

ice

2

kame terraces

outwash plain (*n*) a sheet of sediment (p. 11) laid down by water escaping from a glacier (p. 28) when it melts. Its surface may be cut by a braided river (p. 25).

sandur (*n*) an Icelandic name widely used for an outwash plain (↑). **sandar** (*n.pl.*).

kettle hole (*n*) a hollow in a land surface on drift (p. 33). It results when a block of ice, left behind in the drift, melts. The hollow may contain a small lake.

kettle hole how it may be formed

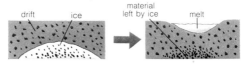

glacial diversion a change in the course of a stream when it is blocked by a glacier (p. 28). The new course is often fixed when the ice melts away.

proglacial lake a body of water that is contained between a glacier (p. 28) and nearby high ground. A new river system may result when the water escapes, as in the English Midlands.

spillway (*n*) a channel cut by meltwater (p. 31) escaping from a proglacial lake (↑). It usually has steep sides and a flat floor, e.g. Newtondale, near Pickering in Yorkshire, England.

proglacial lake and spillway
proglacial lakes and a spillway (Newtondale) in North Yorkshire, England, during the last glacial period

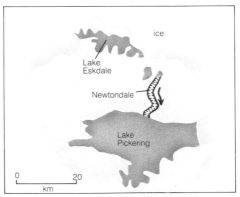

periglacial (adj) of part of the Earth's surface where the freezing and melting of ground ice is the most important process. The word means 'around a glacier' (p. 28) but a glacier is not necessary for periglacial conditions.

freeze-thaw action the action of water when it freezes inside open spaces in rock. The force that results is so high that any rock will break. One of the results is a blockfield (↓).

frost heaving a lifting of the ground that occurs when contained water freezes and so needs more space. Separate clasts (p. 13) may be pushed up, or the ground itself may rise.

solifluction (n) the downhill flow of a wet regolith (p. 19) This process was first noted for periglacial (↑) areas but now includes flow in any areas where slopes are steep and the regolith is wet. It means 'flowing soil'.

névé (n) snow that has lasted through a summer. Melting and pressure has changed the original snow into rounded grains, with glacier (p. 28) ice as the end-product after several years. Also known as **firn**.

nivation (n) the erosive (p. 20) effects that occur under and near a snow bank that stays much the same size. Freeze-thaw action (↑) is the main process and is helped by running water and solifluction (↑).

nivation cirque a large hollow produced by nivation (↑). It may look like a glacial cirque (p. 31), but its floor lacks the over-deepened shape of the true cirque.

permafrost (n) ground which is frozen all the time. It may be very thick, as in Siberia and on the North Slope of Alaska, USA. An active layer (↓) may rest on top.

solifluction

tongues of material resulting from solifluction

névé

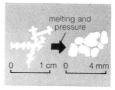

melting and pressure

0   1 cm   0   4 mm

nivation cirque long profile

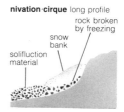

rock broken by freezing
snow bank
solifluction material

permafrost the change in its thickness along a north-south line in Canada

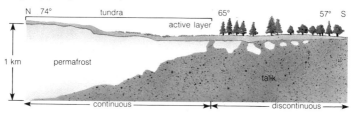

N   74°   tundra   65°   57°   S
active layer
1 km   permafrost
talik
continuous   discontinuous

patterned ground
showing the effect of slope on form

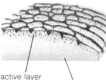

active layer

permafrost

pingo

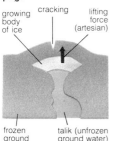

growing body of ice

cracking

lifting force (artesian)

frozen ground

talik (unfrozen ground water)

active layer the surface part of the ground in a periglacial (↑) area whose ice content melts in summer and freezes in winter. As a result the ground may move.

talik (*n*) a layer of material within permafrost (↑) and that has not been frozen. It is usually the place where a pingo (↓) is found.

head (*n*) a periglacial (↑) sediment (p. 11) resting on land surfaces and which was laid down mainly by solifluction (↑). It has usually moved no more than 1 to 2 km, and the rocks it contains point downslope.

patterned ground a set of circles and other regular figures that may be found in the regolith (p. 19) in periglacial (↑) areas. They are formed when ice grows in the ground and so causes material to move about. They may be pulled out into linear forms on slopes.

blockfield (*n*) a sheet of large clasts (p. 13) which result from freeze-thaw action (↑) under periglacial (↑) conditions.

clitter in front of a tor, Dartmoor, England

clitter (*n*) a sheet of clasts (p. 13) of granite (p. 10) that lies on a gentle slope next to a Dartmoor tor (p. 45). It results from freeze-thaw action (↑) during periglacial (↑) times.

pingo (*n*) a small hill with steep sides that results when a body of ice grows inside the ground. The ice may be supplied by water that was already present below or that had flowed under artesian (p. 95) conditions.

thermokarst (*n*) a periglacial (↑) landscape (p. 22) which looks like karst (p. 42) but which results when ground ice melts. Closed hollows, often containing water, are important parts of this landscape.

aeolian (adj) of the process which happens when wind, carrying grains of rock, blows over a land surface. A result may be erosion (p. 20) by abrasion (p. 20), or the laying down of grains to form dunes (↓).

deflation (n) the lowering of a land surface that takes place when loose, dry sediments (p. 11), especially clays (p. 12), are blown away by the wind. It was important in the Great Plains, USA, in the 1930s when much soil was lost.

deflation hollow a large hollow that has resulted from deflation (↑). It may be eroded (p. 20) down as far as the water-table (p. 94) where wind action is not able to remove damp material. Good examples are found in the Sahara and Kalahari deserts.

ventifact (n) a piece of loose rock with gently curved sides. It is a result of abrasion (p. 20) by wind carrying sand grains (p. 13). A ventifact may have several such sides which meet quite sharply. Ventifacts are found in both hot and cold deserts.

yardang (n) a long ridge (p. 43) with a smooth slope that is the result of abrasion (p. 20) by aeolian (↑) activity.

ventifact
ventifacts shaped by wind abrasion

yardang

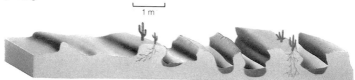

1 m

zeugen (n) a rock which stands above the general surface of the ground and whose lower part has the form of a neck. This shape is the result of aeolian (↑) abrasion (p. 20) and is most clear when the lower part of the rock is soft. **zeuge** (pl).

dune (n) a pile of sand that has been acted on by wind to produce a smooth, regular shape. Several kinds of dune are found in sand deserts (p. 87). They may result from changes in the wind and the amount and size of sand.

zeugen

layer of hard rock

soft rock such as shale

barchan

wind direction

seif dune
how a barchan may grow into
a seif dune

barchan (*n*) a curved dune (↑) that moves slowly
across the surface of a sand desert (p. 87).
Its outer edges move faster than the centre.
The slope facing the wind is quite gentle while
the opposite slope is steeper.

seif dune a quite short sand dune (↑) made up
of curved parts and found in hot deserts (p. 87).
It results when a barchan (↑) dune grows in
length and is built up by winds that blow from
two main directions.

erg (*n*) a large area of sand dunes (↑) especially
in the Sahara, where the Great Eastern Erg
covers about 196,000 km².

erg erg or sand sea

bajada the relation between
bajadas and pediments

mountain front

bajada

pediment

pediment forming against a
steep hill slope

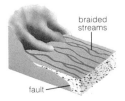

braided
streams

fault

bajada (*n*) a long, gently sloping plain found
against a cliff in a desert (p. 87). It is made
up of many alluvial fans (p. 28) laid down by
streams that only flowed for part of the time.

hamada (*n*) a wide raised land surface in the
Sahara with steep slopes at its borders. It is
often covered by broken pieces of rock.

reg (*n*) a sheet of small rounded pieces of rock
that rests on a nearly flat plain in the Sahara.

pediment (*n*) a land surface with a gentle slope,
up to 5°, that is often found in areas where
the plant cover is slight. It may be separated
from much steeper slopes by a bajada (↑). It
is the result of erosion (p. 20) by small streams
as they move from side to side.

pediplain (*n*) a wide land surface that is made up of many pediments (p. 39).

playa (*n*) a low part of a desert (p. 87) area with high ground around it. As a result it may contain fine material that has been washed in. Water may stand there for some time, and when it dries out salty material is left.

wadi (*n*) *see* arroyo (p. 28).

desertification (*n*) the spreading of a desert (p. 87) into area where crops could once be grown. It may be a result of a change in climate (p. 71), and may be speeded up by too much grazing (p. 138). It is happening in parts of more than 100 countries. It has occurred especially along the edges of the Sahara desert, Africa, and has been made worse by drought (p. 210) conditions.

volcano (*n*) a landform that is built up of material thrown out from some depth in the Earth. It may be made of many kinds of material, including lava (p. 10) and broken volcanic rock (p. 10), and may take many shapes. **volcanic** (*adj*).

shield volcano a volcano (↑) that is made of a magma (p. 10) that flows easily. It often has a caldera (p. 41) at its top. It may be very large: Mauna Loa, in the Hawaiian Islands, is 10,000 m high and 400 km across.

stratovolcano (*n*) a volcano (↑) that is built up of sheets of lava (p. 10) and of volcanic debris (dust and pieces of rock), e.g. Etna.

stratovolcano
its formation.
Note the layers of ash and lava, and the many pipes through which melted rock rises

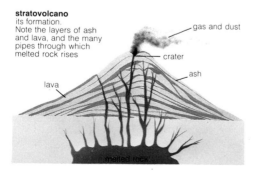

gas and dust

crater

ash

lava

melted rock

planèze on an eroded stratovolcano

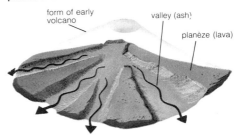

form of early volcano

valley (ash)

planèze (lava)

crater ash cones and craters in the Massif Central, France

simple cone with central crater

additional cones on the side

cone-in-cone forms

a lava flow has cut through one side

planèze (n) part of the surface of a stratovolcano (↑) made up of a sloping sheet of lava (p. 10). It is often cut by quite deep valleys, and may stand well above the general surface of the volcano (↑). Good examples are found on Le Cantal, Massif Central, France.

crater (n) a small hollow made by the activity of a volcano (↑). It is a result of explosion, e.g. at the top of an ash cone, or below a flat surface when a maar (↓) results.

caldera (n) a large hollow at the top of a volcano (↑). It comes about when the top of the volcano sinks, but explosion may also play a part. The caldera of Mt Teide in Tenerife is up to 17 km wide.

maar (n) a nearly round hollow, often filled with water, and which results from a volcanic (↑) explosion. This may happen because groundwater (p. 94) meets very hot rock at depth. **maare** (pl).

caldera
Las Cañadas, Mount Teide, Tenerife. Teide itself was built up later

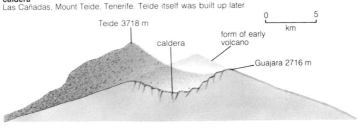

Teide 3718 m

caldera

form of early volcano

Guajara 2716 m

0   5
km

karst (*n*) the landscape (p. 22) produced by the weathering (p. 19) and erosion (p. 20) of the hard limestone (p. 12) area of Yugoslavia. It is now used to describe the landforms of any limestone area. **karstic** (*adj*).

karren (*n*) the group of forms that result from weathering (p. 19) on the surface of a hard limestone (p. 12). The group includes grikes (↓).

clint (*n*) a small limestone (p. 12) surface about 1 to 2 m across, whose borders are steep joints (p. 14). Its surface is part of a bedding plane (p. 18) and may snow small hollows that result from weathering (p. 19). A landform found in the north Pennine Hills, England.

grike (*n*) a joint (p. 14) in limestone (p. 12) that has been widened by weathering (p. 19); especially at the border of a clint (↑). It may be up to 0.6 m wide and 3.0 m deep in the north Pennine Hills, England.

limestone pavement a level area where hard limestone (p. 12) is found at the surface. It is a former bedding plane (p. 18) from which erosion (p. 20) has removed the rock above. It may be made up of clints (↑) and grikes (↑).

doline (*n*) a hollow with steep sides that is found in a limestone (p. 12) area. It is a result of weathering (p. 19) especially at places where joints (p. 14) are close together. A natural tube may lead from it to a cave below.

polje (*n*) a large hollow with steep sides and a flat floor that is found in a karst (↑) area. It may result from a fault (p. 14) or from weathering (p. 19). An important landform in the karst of Yugoslavia.

dry valley a valley cut by a stream through a limestone (p. 12) area, but which no longer contains water. Many ideas have been put forward to explain the loss of water, e.g. a fall in the water-table (p. 94) or an increase in the infiltration capacity (p. 96) of the rock.

inverted relief a landscape (p. 22) or landform whose shape is opposite to the way the rock structure (p. 14) below is arranged. For example, an anticlinal valley (↓) and a synclinal ridge.

limestone pavement
flat, wide clints and narrow grikes on a limestone pavement

clint   grike

doline
doline produced by weathering

clay

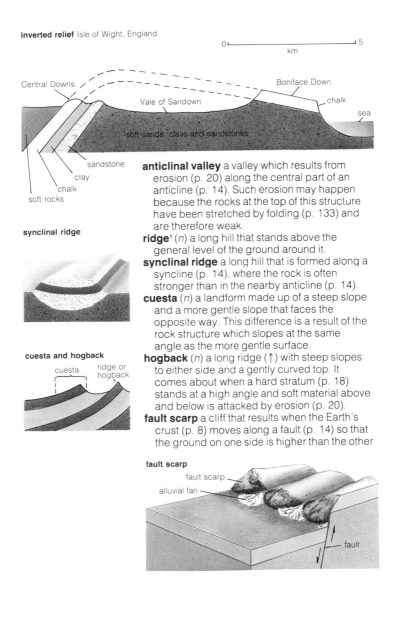

inverted relief Isle of Wight, England

synclinal ridge

cuesta and hogback

fault scarp

anticlinal valley a valley which results from erosion (p. 20) along the central part of an anticline (p. 14). Such erosion may happen because the rocks at the top of this structure have been stretched by folding (p. 133) and are therefore weak.

ridge[1] (*n*) a long hill that stands above the general level of the ground around it.

synclinal ridge a long hill that is formed along a syncline (p. 14), where the rock is often stronger than in the nearby anticline (p. 14).

cuesta (*n*) a landform made up of a steep slope and a more gentle slope that faces the opposite way. This difference is a result of the rock structure which slopes at the same angle as the more gentle surface.

hogback (*n*) a long ridge (↑) with steep slopes to either side and a gently curved top. It comes about when a hard stratum (p. 18) stands at a high angle and soft material above and below is attacked by erosion (p. 20).

fault scarp a cliff that results when the Earth's crust (p. 8) moves along a fault (p. 14) so that the ground on one side is higher than the other

rift valley the Gregory Rift, East Africa

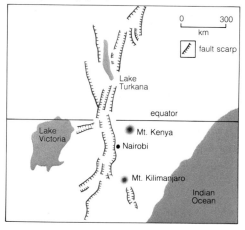

rift valley a valley that is formed when part of the Earth's surface is stretched. The land sinks between the two faults (p. 14) that result, and volcanoes (p. 40) may form, as in the Gregory Rift, Kenya. Such a valley is larger than one that results from erosion (p. 20) alone.

bornhardt (*n*) a large round hill that may stand on its own and is often made of granite (p. 10) or a granite-like rock. Its round shape results from curved sheets of rock that are separating from the solid rock below.

inselberg (*n*) an 'island mountain' which stands well above a plain, often because it is made of harder rock than the country around it. **inselberge** (*n.pl.*).

bornhardt

bornhardt

inselberg

inselberg with a cap of hard rock

tor

tor (*n*) a hill of bare rock about the size of a house, which is found especially on granite (p. 10). Its shape is a result of the way in which the joints (p. 14) are arranged. On Dartmoor, England, it may have a sheet of clitter (p. 37) around it.

butte (*n*) a small hill with a flat top that stands on its own. Its shape is a result of the erosion (p. 20) of rock strata (p. 18) that are nearly level.

butte and mesa

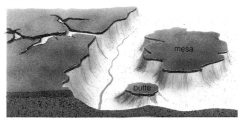

mesa (*n*) a hill with a wide flat top that stands on its own. It is similar to a butte (↑) but its width is much greater than its height.

antecedent stream a stream that is older than a rock structure it flows across. Such a stream cut down its channel as fast as the structure was formed. The Grand Canyon of the Colorado river in the USA is antecedent.

antecedent stream

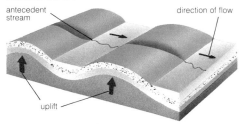

antecedent stream

direction of flow

uplift

subsequent stream a stream which follows a path where rocks are weak, often because of a fault (p. 14), a main joint (p. 14) or the outcrop (p. 18) of a soft stratum (p. 18)

raised beach

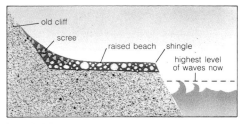

- old cliff
- scree
- raised beach
- shingle
- highest level of waves now

beach (*n*) a pile of loose sand or of larger pieces of rounded rock called shingle, and which lies at the edge of the sea. Its form changes rapidly as outside conditions (e.g. wave activity) change, and so it is an example of dynamic equilibrium (p. 220).

shingle (*n*) *see* beach (↑).

raised beach a beach (↑) which now stands well above sea-level. This may be because the sea-level has fallen since the beach was laid down, perhaps as a result of the growth of ice-sheets, or because the land has risen.

berm (*n*) a ridge (p. 43) with a flat top that lies at the highest line on a beach (↑) that waves have reached.

wave refraction the bending of a wave in plan as seen from above as it passes through water of lessening depth. This happens because wave speed slows down as water becomes less deep. So, that part of the wave still in deep water moves at its starting speed while the rest slows down, so bending the wave.

swash (*n*) the water that rushes up a beach when a wave breaks. It is most important when the backwash (↓) of the wave before has already returned to the sea and so does not hold up the movement of the swash. It may carry beach material towards the land.

backwash (*n*) the water that flows back down a beach (↑) after the swash (↑) has reached its highest point. The amount is controlled by the infiltration capacity (p. 96) of the beach, and by the swash of the next wave. It may move beach material towards the sea.

wave refraction the wave is bent in plan as it moves into less deep water, and its direction changes from **1** to **3**

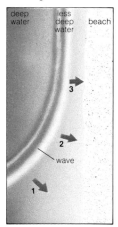

deep water
less deep water
beach

3

2

wave

1

beach drift

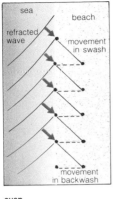

cusp

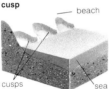

beach drift the movement of sediment (p. 11) along a beach (↑). It takes place along the swash (↑) – backwash (↑) area of wave activity, and is most effective when waves arrive at a sharp angle.

longshore drift the general movement of material along a coast. It takes place both by beach drift (↑) and by movement, caused mainly by large waves.

cusp (*n*) a small beach (↑) form. It is made up of two horns of relatively large beach material, separated by a lower area of fine sand. It ranges in size from 1 m to 60 m and may be formed by the swash (↑) and backwash (↑). **cuspate** (*adj*).

groyne (*n*) a fence made of wood or metal that runs down a beach (↑) and enters the sea. Its purpose is to trap sediment (p. 11) moving along the beach by longshore drift (↑) and so prevent erosion (p. 20).

spit (*n*) a beach (↑) that leaves the shore, often at a point where the coast curves inland, and goes into deeper water. Its end may have recurves (↓).

spit Hurst Castle spit, England, with recurves

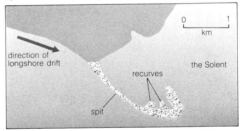

recurve (*n*) a low ridge (p. 43) of sand or shingle (↑), shaped in plan like a hook, and which joins the outer end of a spit (↑). A smoothly rounded recurve may result from wave refraction (↑) as the end of the spit grows into deeper water. An angular recurve may result from waves arriving from a direction other than that which controls the main spit.

tombolo (*n*) a beach (p. 46) that ties an island to the coast near by. Portland Bill, in southern England, is tied to the coast by a tombolo.

offshore bar a ridge (p. 43) of sediment (p. 11) that lies up to 40 km out to sea. It is a result of waves breaking along a line which stays in nearly the same place because the daily change of sea-level is small.

estuary (*n*) the lower part of a river valley that is now covered by water, usually as a result of the rise of sea-level that followed the last glacial period (p. 30). It may contain much fine-grained sediment (p. 11) on which salt marshes (↓) are found. **estuarine** (*adj*).

salt marsh a flat area on part of the coast sheltered from wave action and where mud has been laid down. It shows a history of changing plant cover as time passes.

coastal dune a heap of sand found usually in the area of country between the edge of the sea and up to about 10 km inland. For a coastal dune to form, there must be strong winds from the sea, much sand, and a plant cover. Various kinds of dune are found.

tombolo near Nelson, South Island, New Zealand

dune slack
relation between dune slacks and the water-table

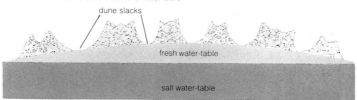

dune slack a low wet area between coastal dunes (↑). It is a result of wind erosion (p. 20) down to the water-table (p. 94) and has a special plant community (p. 73).

hydraulic action the direct attack by waves on solid rocks. It results from the large forces set up when waves break against rock, and especially when air is forced suddenly into joints (p. 14). It is the most important way by which waves erode (p. 20) cliffs.

stack

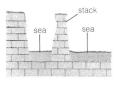

notch (*n*) an under-cut form at the foot of a sea cliff. As it gets bigger through wave action, the rocks above fall, and so the cliff moves inland.

visor (*n*) that part of a sea cliff above the notch (↑) and which stands well forward. It is formed especially on limestone (p. 12) coasts in the tropics (p. 241) where sea-level changes little and chemical weathering (p. 19) is important.

stack (*n*) a finger of rock that often stands close to a sea cliff. It is a result of erosion (p. 20) along steep joints (p. 14).

geo (*n*) a long, narrow, and steep-sided valley that runs back into a sea cliff and which may follow an important joint (p. 14). It may have resulted when the roof of a cave fell in.

strandflat Norway

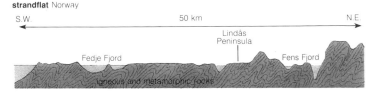

ria the coast of southwest Ireland

peninsula

strandflat (*n*) a wide area of the coast of Norway, consisting of low islands and a gently sloping surface just below sea-level. It is cut across hard rocks, but the way it was produced is not yet understood.

shore platform a nearly flat area of bare rock that is found just off shore and close to mean (p. 226) sea-level. It is cut by the action of waves and weathering (p. 19).

fiord (*n*) a steep-sided valley which is the result of erosion (p. 20) by a glacier (p. 28) and which is now below sea-level. Many examples are found in Norway.

ria (*n*) a valley that was produced by river and hill slope erosion (p. 20) and then disappeared as the sea-level rose. Many examples, as in southwest Ireland, resulted from the rise in sea-level as the last ice sheets (p. 29) melted.

peninsula (*n*) a piece of land that is bordered by water on three sides. **peninsular** (*adj*).

Atlantic coast a coast that cuts across the geological structures of an area at a high angle. An example is the coast of southwest Ireland. Also known as **transverse coast**.
transverse coast = Atlantic coast (↑)
Pacific coast a coast that runs in the same direction as the geological structures of the area. An example is the Dalmatian coast of Yugoslavia.

guyot flat-topped shape of a typical guyot

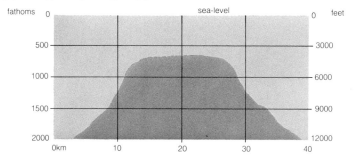

guyot (*n*) a volcanic (p. 40) hill with a flat top beneath the sea. Cut by wave action, it may now be between 400 m and 200 m below sea-level. About 2000 have been found.
seamount (*n*) a volcanic (p. 40) hill beneath the sea. It still has its former shape. A guyot (↑) is a type of seamount.
atoll (*n*) a ring-shaped form, made of limestone (p. 12) produced by very small sea creatures, and with open water in the centre. Many atolls in the open ocean rest on sediments (p. 11) laid down when the sea was less deep showing that the ocean floor there has sunk.
trench (*n*) a long, narrow and deep part of the ocean floor. It may lie near a land mass, e.g. the Peru-Chile trench off the west coast of S. America, or it may be found near an island arc (p. 16), e.g. the Marianas trench of the West Pacific, which is the world's deepest, at 11,022 m below sea-level. A trench is formed along the line of a descending plate (p. 14).

atmosphere   atmospheric temperature and pressure

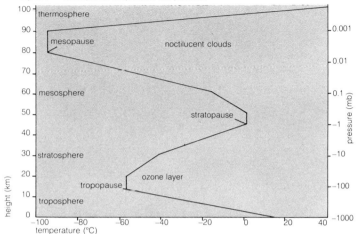

atmosphere (*n*) the gaseous body around the Earth. **atmospheric** (*adj*).

troposphere (*n*) the lowest part of the atmosphere, where weather (p. 67) activity is most marked. The troposphere has 75% of the gaseous mass of the atmosphere. There is a mean (p. 226) temperature fall of 6.5°C per kilometre of height up to the tropopause (↓). **tropospheric** (*adj*).

tropopause (*n*) the top of the troposphere (↑), where there is a general inversion (p. 54) or an isothermal layer (p. 52). The tropopause has a mean (p. 226) height that ranges from 16 km at the equator (p. 238) to 8 km at the poles (p. 239).

stratosphere (*n*) the part of the atmosphere above the tropopause (↑) in which temperature increases with height (up to the stratopause), because of radiation (p. 52) trapped by the ozone layer (p. 52). **stratospheric** (*adj*).

stratopause (*n*) the top of the stratosphere (↑) where the increase in temperature which characterizes the stratosphere stops. It has a mean (p. 226) height of 50 km.

ozone layer the part of the stratosphere (p. 51) between 20 and 25 km above the Earth which is relatively rich in the gas ozone (O_3).

mesosphere (*n*) the part of the atmosphere above the stratopause (p. 51), where the temperature falls continuously with height to about $-90°C$ at around 80 km, i.e. at the mesopause ($\downarrow$). **mesospheric** (*adj*).

mesopause (*n*) the inversion (p. 54) at the top of the mesosphere ($\uparrow$) where temperatures begin to rise.

thermosphere (*n*) the part of the atmosphere above the mesopause ($\uparrow$). The amount of gas is very small and although temperature is described as increasing with height, the increase is slight as there is so little gas. **thermospheric** (*adj*).

ionosphere (*n*) the lower part of the thermosphere ($\uparrow$), particularly between heights of about 100 km and 300 km where there are relatively large numbers of ions, atoms or molecules produced by various kinds of radiation ($\downarrow$). **ionospheric** (*adj*).

magnetosphere (*n*) the outer part of the atmosphere, which consists of very small amounts of electrically charged matter from the Sun. The trapping of this matter in the Earth's atmosphere gives rise to the so-called Northern (*aurora borealis*) and Southern Lights (*aurora australis*). **magnetospheric** (*adj*)

isothermal layer a part of the atmosphere where there is no change of temperature with height.

radiation (*n*) the movement of energy (both heat and light) from one body to another, at the speed of light, and without the need for any matter in between, i.e. it can cross empty space. **radiate** (*v*), **radiator** (*n*).

conduction (*n*) the passage of heat through a substance, from point to point.

convection (*n*) the upward movement of heat in masses of air. A body of air which is warmer than the air around it, will rise because it is less heavy. **convective** (*adj*).

advection (*n*) the movement of air, energy, etc, across rather than up from a flat surface.

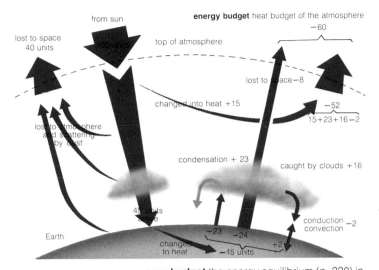

energy budget heat budget of the atmosphere

from sun

lost to space
40 units

top of atmosphere

−60

100 units

lost to space −8

changed into heat +15

−52
15+23+16−2

lost to atmosphere
and scattering
by dust

condensation + 23

caught by clouds +16

45 units
arrive

conduction −2
convection

Earth

−23 −24
+2

changed
to heat −45 units

greenhouse effect

energy

atmosphere
lets through
shortwave
radiation
from sun

atmosphere traps much outgoing
infrared radiation from Earth

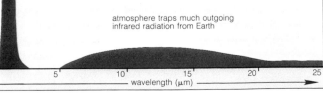

0   5   10   15   20   25
wavelength (μm)

energy budget the energy equilibrium (p. 220) in
the atmosphere. The amount of energy gained
by the atmosphere is exactly the same as that
which is lost over a year, and so there is no
general increase or decrease of temperature,
other than at times of climatic (p. 71) change.
Within this overall equilibrium, the budget
consists of the various kinds of energy gains
and losses. Also known as **energy balance**.

greenhouse effect the increase in temperature
of the lower part of the atmosphere, due to the
trapping of incoming radiation (↑) by the gas
carbon dioxide and by water in gaseous form,
i.e. water vapour. This effect is believed to be
increasing as a result of air pollution (p. 197).

albedo (*n*) the amount, expressed as a percentage, of radiation (p. 52) falling on a surface which is lost, i.e. reflected back into space. Light surfaces have higher albedo values than dark ones. The average albedo of the Earth and atmosphere is 31% of the incoming radiation from the Sun.

inversion (*n*) an increase of something, e.g. temperature with height, instead of the usual fall.

isotherm (*n*) a line of equal temperature.

adiabatic (adj) of a change in an atmospheric characteristic (e.g. temperature, pressure) when there is no heat added to or taken from the body of air concerned. Thus when a mass of air increases in size as a result of a fall in pressure, there is also a fall in temperature.

humidity (*n*) the amount of gaseous water, i.e. water vapour, in the atmosphere. Absolute humidity is the actual amount present and is measured in grams per cubic metre (g m^{-3}). *Relative humidity* expresses the actual water content of a volume of air as a percentage of the full amount of water that the same volume of air could hold at the same temperature.

condensation (*n*) the change from gas to liquid.

condensation nucleii microscopic, wettable substances which serve as centres of condensation (↑). *See* aerosol (p. 200).

Bergeron-Findeisen theory a theory (p. 222) which states that precipitation (↓) forms as a result of the freezing of supercool (↓) water direct onto freezing nucleii (↓), which then grow to the point where they are heavy enough to fall. Without such nucleii small water droplets can be supercooled to −40°C, but with them freezing begins at −10° to −15°C.

supercooling in relation to water, the lowering of the temperature below the freezing point of water without ice forming. **supercool** (*adj, v*).

freezing nucleii the particular kinds of microscopic aerosols (p. 200), less common than condensation nucleii (↑), which allow supercool (↑) water to freeze at much higher temperatures than would be the case without them.

stability
stable and unstable air
a stable
uplifted air cooler, heavier
and so tends to sink

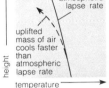

b unstable
uplifted air warmer, lighter
and so tends to rise

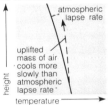

collision-coalescence theory the theory (p. 222)
that relates to raindrop formation in 'warm'
clouds, when large water droplets fall faster
than small ones, thus catching them up and
joining with them. This continues to the point
where the droplet is large enough to fall.

fog (n) the condensation (↑) of water at ground
level, such that objects at a distance of 1 km
or more cannot be seen. If objects beyond
this can still be seen, it is called mist.

stability (n) a condition in which a small change
in a system (p. 217) at equilibrium (p. 220)
produces forces that tend to return the system
to equilibrium. **stable** (adj).

instability (n) a condition in which a small change
in a system (p. 217) at equilibrium (p. 220)
produces forces that tend to move the system
farther away from equilibrium. **unstable** (adj).

conditional instability if an air mass (p. 63)
which appears to be stable is forced to rise,
e.g. when passing over a mountain, and then
continues rising, the original state is conditional
instability. It is conditional on the amount of
condensation (↑) that takes place after uplift
(rising). Condensation slows down cooling,
and the warmer the air mass the lighter it is.

lapse rate the amount of decrease of temperature
with height in the atmosphere. When there is
no condensation (↑) taking place it is called
the *dry adiabatic* (↑) *lapse rate*. This lapse rate
is always the same at 9.8°C km⁻¹. Uplift (rising)
often produces condensation, and as this
slows the speed of cooling, the *saturated
adiabatic lapse rate* is smaller than the dry
one. It varies with temperature, between
4°C km⁻¹ at high temperatures to 9°C km⁻¹ at
−40°C. If air descends it will warm up at one or
other of the above lapse rates. *See* instability (↑).

orographic effect the effect that mountains have
on the behaviour of the atmosphere, especially
on the formation of clouds and precipitation (↓).

precipitation (n) any form of liquid or solid water
that falls from the atmosphere to the ground,
e.g. rain, snow, sleet (rain and snow) and hail
(rounded or irregular pieces of ice).

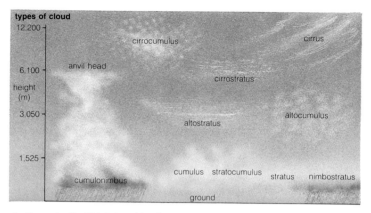

types of cloud

12,200 —

cirrocumulus   cirrus

6,100 —   anvil head

                                    cirrostratus

height
(m)

3,050 —   altocumulus

                        altostratus

1,525 —

            cumulus   stratocumulus   stratus   nimbostratus

cumulonimbus

ground

cirriform (adj) of high-level feather cloud
consisting of threads of ice grains. Cirrus and
the various other cirriform clouds form above
a height of 3 to 6 km, depending on latitude
(p. 238). The name cirrus is used with those of
the other two main cloud shapes – cumulus
(↓) and stratus (↓) – for clouds with shared
characteristics, e.g. cirrostratus. **cirrus** (n).

cumuliform (adj) of cloud with a heaped
appearance showing steady upward growth.
As well as cumulus, the cumuliform clouds
include: cirrocumulus (*see* cirrus (↑));
altocumulus (*see* alto-((↓)); stratocumulus
(*see* stratiform (↓)) and cumulonimbus (↓).
cumulus (n).

stratiform² (adj) of layer-shaped cloud. As well
as stratus, the stratiform clouds include:
cirrostratus (*see* cirrus (↑)); altostratus (*see*
alto-(↓)); nimbostratus (*see* nimbus (↓)). and
stratocumulus (*see* cumulus (↑)). **stratus** (n).

alto (*adj*) of middle-level cloud, usually below
the height of the cirriform (↑) group and above
2 km. There are two sorts of middle-level
cloud, namely altocumulus (*see* cumulus (↑))
and altostratus (*see* stratus (↑)).

nimbus (n) thick cloud from which continuous
precipitation (p. 55) is falling, i.e. nimbostratus
(*see* stratus (↑)) and cumulonimbus (↓).

cumulonimbus (*n*) a large cumuliform (↑) cloud which flattens outwards at the top to give a so-called anvil head because of its shape.

noctilucent clouds very high-level clouds which form at the mesopause (p. 52) and which are seen over high latitudes (p. 238) in summer. The clouds consist of ice grains in which the freezing nucleii (p. 54) appear to be dust from space.

lenticular cloud formation of lenticular and rotor clouds

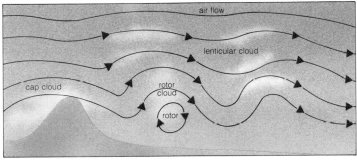

anabatic and katabatic valley winds

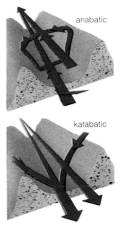

lenticular cloud the cloud which forms when condensation (p. 54) takes place at the top of a wave or waves which result from air flow over a hill or mountain.

rotor cloud a cloud formed when air passes over a hill or mountain, and the air in the lower part of the wave or waves nearest to the hill or mountain moves in a circular manner. The uplift this causes results in condensation (p. 54) and rotor cloud formation.

anabatic (adj) of winds or breezes (i.e. light winds) which blow up-slope in valleys. Valley sides are heated by day and the air resting on them becomes warmer and lighter, so that it rises. Anabatic winds are strongest at about 1400 hours when the Sun is at its hottest.

katabatic (adj) of winds which flow down valley sides and valley floors. At night valley slopes cool and the air resting on them also cools and becomes heavier when it moves down the slope.

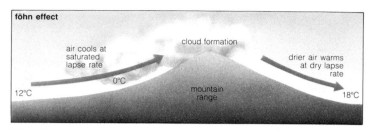

föhn effect

air cools at saturated lapse rate

cloud formation

drier air warms at dry lapse rate

0°C

12°C

mountain range

18°C

föhn effect the warming of some airstreams as they descend a mountain range. This can happen when: (1) air which rises over mountains may experience condensation (p. 54) and precipitation (p. 55) and thus cool slowly at the saturated adiabatic lapse rate (*see* lapse rate (p. 55)). On descending the mountains the drier air warms faster at the dry adiabatic lapse rate; (2) a temperature inversion (p. 54) at mountain-top level blocks the air flow across a mountain range. This forces air from higher levels to descend and warms adiabatically (p. 54). The temperature rise in either (1) or (2) can be rapid, with sudden melting of snow and ice. Where the effect is very marked the wind is given a regional (p. 241) name, e.g. Föhn in the Alps and Chinook in the Rocky Mountains.

fall wind (*n*) a cold wind descending a mountain range. It happens when a cold polar (p. 239) continental air mass (p. 63) is drawn across a mountain range by the pressure gradient (↓). Although adiabatic (p. 54) warming takes place, the air mass is still colder than the one which it replaces. For example, the Bora of the northern Adriatic.

land and sea breezes light winds which blow off the land (land breeze) or off the sea (sea breeze). By day the land becomes warmer than the sea so that the air over the land is warmer and thus lighter than the air over the sea. Thus cooler, heavier air blows inland from off the sea. The opposite effect takes place at night, when the land becomes cooler than the sea because it loses heat more quickly.

land and sea breezes

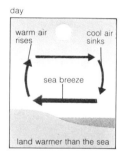

day

warm air rises

cool air sinks

sea breeze

land warmer than the sea

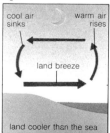

night

cool air sinks

warm air rises

land breeze

land cooler than the sea

harmattan (*n*) a hot, dry wind which blows out of North Africa towards the Mediterranean. The air is from the Hadley cell (p. 62).

mistral (*n*) the cold, northerly winds of the Gulf of Lions, caused when a depression (p. 66) in the Gulf of Genoa draws in colder air from the north. The same winds are known as the Tramontana in Spain. Katabatic (p. 57) effects and the very form of large valleys add to the strength of the mistral, especially in the Rhône Valley.

scirocco (*n*) a hot, dry, dusty wind or airstream of Algeria and the Levant. The air mass characteristics are tropical continental (*see* air mass classification (p. 64)), and the air is drawn in towards the Mediterranean by eastward-travelling depressions (p. 66). The scirocco usually blows in spring and autumn, when sea breezes (*see* land and sea breezes (↑)) are less than in summer. The same winds are called leveche in southeast Spain and khamsin in Egypt.

trade wind the tropical (p. 241) easterly winds which blow in from either side of the equator (p. 238) towards the ITCZ (p. 70) from the subtropical high pressure areas formed where air descends from the Hadley cells (p. 62). Trade winds have a direct effect on the surface currents of the oceans.

pressure (*n*) the force acting on a given area. Atmospheric pressure depends on the weight of the atmosphere above, i.e. resting on, the given area; it is measured in millibars (mb), 1 mb being equal to a force of 100 newtons acting on $1m^2$ ($1Nm^{-2}$). The Nm^{-2} can also be called Pa (pascal).

isobar (*n*) a line joining points of equal pressure.

pressure gradient the difference in pressure between various places. Wind blows from areas of high pressure towards areas of lower pressure. The larger the difference the greater the pressure gradient force and thus the stronger the wind. Other forces prevent the air moving directly across the isobars (↑). *See* geostrophic wind (p. 60).

coriolis force a force produced by the turning of the Earth in space, which tends to bend the path of objects moving relative to the Earth's surface, to the right in the northern hemisphere (p. 238) and to the left in the southern.

angular momentum an effect produced by the tendency of the Earth's atmosphere to move with the Earth as it turns in space. Angular momentum is greatest at the equator (p. 238) and disappears at the poles (p. 239). The overall angular momentum of the atmosphere must stay the same, i.e. there is a conservation of angular momentum. Thus air moving from one latitude (p. 238) to another must also experience a change in its speed or velocity, e.g. air moving polewards will tend to move more quickly. The same kind of thing happens when a person turning quickly on the same spot on ice pulls his or her arms into the body. In practice, increased air speeds due to angular momentum are partly offset by other forces; even so, the poleward movement of angular momentum is an important part of the general circulation (↓) of the atmosphere.

centripetal acceleration the change in direction of movement when a body moves in a curved path around a centre, e.g. of high pressure.

centrifugal force[1] a force, equal but opposite to centripetal acceleration (↑), which appears to try to bend a body moving along a curved path, outwards from the centre of curvature.

geostrophic wind a wind which blows more or less at right angles to the pressure gradient (p. 59) (i.e. along rather than across the isobars (p. 59)), when the coriolis force (↑) exactly equals the pressure gradient force, acting in the opposite direction to it.

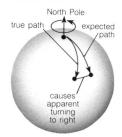

coriolis force

North Pole

true path

expected path

causes apparent turning to right

geostrophic wind

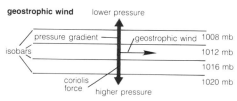

lower pressure

pressure gradient — geostrophic wind 1008 mb

isobars 1012 mb

1016 mb

coriolis force 1020 mb

higher pressure

gradient wind

1 low pressure

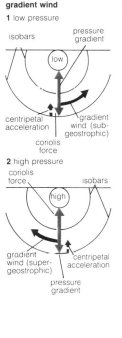

isobars

pressure gradient

low

centripetal acceleration

gradient wind (sub-geostrophic)

coriolis force

2 high pressure

coriolis force

isobars

high

gradient wind (super-geostrophic)

centripetal acceleration

pressure gradient

gradient wind (*n*) the steady flow of air along a curved path, e.g. in low and high pressure systems (p. 217). For this to happen there must be a difference between the coriolis force (↑) and the pressure gradient (p. 59) force to produce the necessary centripetal acceleration (↑) for movement along a curved path. In low pressure systems the coriolis force is the smaller, so that wind speeds are less than those for geostrophic winds (↑), i.e. they are subgeostrophic. In high pressure systems the opposite is true and wind speeds are super-geostrophic. As low pressure systems generally have stronger pressure gradients than high pressure systems, these effects are not obvious.

veer (*v*) of winds, a clockwise change in direction.

back (*v*) of winds, an anticlockwise change in direction.

gust (*n*) a sudden but short increase in wind speed. **gusty** (*adj*).

general circulation the general movement of the Earth's atmosphere, giving characteristic systems (p. 217) of wind and pressure that are either seasonal or all-year-round in nature.

general circulation
generalized wind and pressure belts of the world

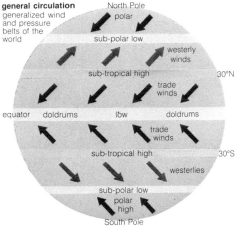

North Pole

polar

sub-polar low

westerly winds

sub-tropical high   30°N

trade winds

equator   doldrums   low   doldrums

trade winds

sub-tropical high   30°S

westerlies

sub-polar low

polar high

South Pole

Hadley cell longitudinal atmospheric circulation model for the northern hemisphere

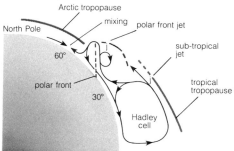

Hadley cell an idealized system (p. 217) in which there is an uplift of warm air in great cumulonimbus (p. 57) clouds along the ITCZ (p. 70), and the sinking of this air at about 30° north and south of the equator (p. 238) into the high pressure areas of the so-called 'horse latitudes' (↓). The surface flows from these high pressure areas back to the ITCZ are deflected by the coriolis force (p. 60) to give the easterly trade winds (p. 59).

horse latitudes areas of high pressure and little or no wind about 30° north and south of the equator (p. 238). *See* Hadley cell (↑).

jet stream a narrow band of very strong geostrophic winds (p. 60) within the upper westerlies (↓). Speeds of 160–240 k.p.h. are common, increasing to as much as 480 k.p.h. in winter. Jet streams are found where sharp temperature changes produce very strong pressure gradients (p. 59) over short distances. Thus the polar front jet stream blows above the polar front (p. 64), and the sub-tropical jet stream blows on the poleward side of the Hadley cells (↑), where there is often a marked temperature difference between the sinking tropical (p. 241) air and the air of the middle latitudes (p. 238).

prevailing westerlies wind patterns that blow west to east around latitudes 40°N and 40°S of the equator (p. 238).

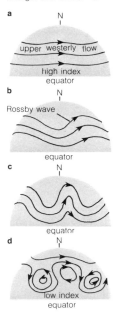

zonal index cycle
changes over time **a – d**

a

N

upper westerly flow

high index
equator

b

N

Rossby wave

equator

c

N

equator

d

N

low index
equator

upper westerlies the upper tropospheric (p. 51) winds that are present over most of the Earth. The height of a particular pressure surface, e.g. the 500 mb surface, is largely controlled by atmospheric temperature, the warmer the air the higher the surface. Thus the 500 mb surface is much higher towards the equator (p. 238) than it is nearer the poles (p. 239). This gives a marked pressure gradient (p. 59) in the upper troposphere of each hemisphere (p. 238), with the 'lows' centred over the poles. Geostrophic winds (p. 60) result, which blow around the world from west to east. *See* jet stream (↑).

circumpolar westerlies = upper westerlies (↑).

Rossby waves long waves in the upper westerlies (↑). Because of their large size, only between three and six such waves can be contained within the upper westerly air flow of each hemisphere (p. 238). They are more or less fixed in place, and in overall spatial plan seem to relate closely to great mountain ranges and marked changes in land/sea-surface temperatures. The Rossby waves largely control the track taken by depressions (p. 66).

zonal index cycle the growth of irregular flow in the upper westerlies (↑) from a condition of high zonal index to one of low zonal index. When the index is high the westerlies are strong, waves in the system (p. 217) are gentle and there is little north-south movement of air. Then the waves grow steadily, and, towards the close of the cycle, when the index is low, the westerlies break up into a number of more or less fixed cold depressions (p. 66) in lower middle latitudes (p. 238) and deep warm blocking anticyclones (p. 67) at higher latitudes.

air mass a large body of air in which such characteristics as temperature, humidity (p. 54) and lapse rate (p. 55) are more or less the same over great geographical areas.

barotropic (adj) of an atmosphere in which surfaces of constant pressure do not cross surfaces of constant density. Thus, up to the tropopause (p. 51), isobars (p. 59) and isotherms (p. 54) stay parallel to each other and the Earth's surface.

baroclinic (adj) of an atmosphere in which surfaces of constant (unchanging) pressure cross surfaces of constant density. Unlike barotropic (p. 63) air, baroclinic air cannot travel as an unchanging, solid body because of its more variable character.

air mass classification air masses (p. 63) are classified according to: (1) temperature – arctic (p. 239), polar (p. 239) and tropical (p. 241) air; and (2) whether the starting point was over land or sea – continental or maritime (↓) air. Thus polar maritime air is relatively cold and damp, while tropical continental air is relatively warm and dry.

maritime (adj) of the sea.

air mass modification the changes which happen when air masses (p. 63) move away from the areas of formation across surfaces of different character. Temperature and humidity (p. 54) changes take place and barotropic (p. 63) conditions give way to baroclinic (↑) ones.

front (n) usually regarded as the sloping surface between two air masses (p. 63), and shown as a line where it meets the ground. In reality a front is a narrow layer of air, usually with rapid changes in such characteristics as temperature, humidity (p. 54) and density. **frontal** (adj).

frontogenesis (n) the formation of a front (↑). Fronts suddenly appear, become wave-like in plan and bigger, and then slowly disappear.

frontolysis (n) the decay of a front (↑).

arctic front the front (↑) between arctic (p. 239) and polar (p. 239) air masses (p. 63).

polar front the front between polar (p. 239) and sub-polar air masses (p. 63) on one hand, and tropical (p. 241) and sub-tropical on the other.

cold front a front (↑) where the colder air mass (p. 63) follows the warmer one. Usually cold fronts travel behind the warm sector (↓) of a depression (p. 66).

warm front a front (↑) where the warmer air mass (p. 63) follows the colder one, usually along the leading edge of a warm sector (↓) of a depression (p. 66).

air mass classification
air masses affecting London: thickness of arrow shows relatively how often the air mass affects London

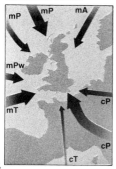

m = maritime
c = continental
P = polar
A = arctic
T = tropical
w = warmer due to more southerly track

occlusion

warm occlusion

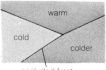

occluded front

cold occlusion

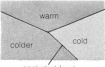

occluded front

ana-front fronts (↑) where air in the warm sector (↓) rises relative to the cold front (↑) and warm front (↑). Ana-fronts are usually very active.

kata-fronts fronts (↑) where the air of a warm sector (↓) sinks relative to the cold air. The cold front (↑) and warm front (↑) are relatively weak.

occlusion (*n*) an event where the cold front (↑) catches up with the warm front (↑) of a depression (p. 66) and lifts the warm sector (↓) off the ground. There are two kinds of occlusion. A *cold occlusion* forms when the air behind the warm sector is colder than the air in front of it, while the opposite produces a *warm occlusion*.

occluded front a front (↑) which marks the line of an occlusion (↑).

warm sector the area of warm air within the wave of a frontal (↑) depression (p. 66).

convergence and divergence

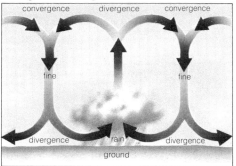

convergence[1] (*n*) air flow into an area. Provided there is no density change at the level at which the convergence takes place, the incoming air must rise.

divergence[1] (*n*) air flow out of an area. Divergence near the ground results in atmospheric sinking or subsidence, provided there are no density changes at the level of the divergence.

depression a life history

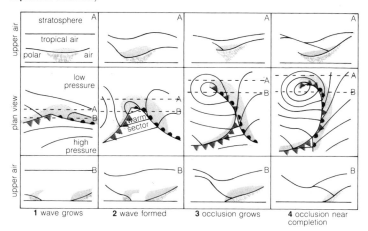

upper air | plan view | upper air

1 wave grows   2 wave formed   3 occlusion grows   4 occlusion near completion

depression (*n*) an area of low pressure, with more or less circular isobars (p. 59). In mid-latitudes (p. 238) cyclogenesis (↓) usually involves convergence (p. 65) of different air masses (p. 63), the front (p. 64) between them growing into a wave. In plan, the top of the wave lies at the centre of the depression. The surface convergence of air masses is caused by divergence (p. 65) in the upper westerlies (p. 63), when the gradient wind (p. 61) changes from sub-geostrophic (p. 60) in a trough (↓) to super-geostrophic in the next ridge (↓). It is the super-geostrophic flow which leads to the upper air divergence. The pressure gradient (p. 59) of a depression is towards the centre, so that the gradient wind blows anticlockwise in the northern hemisphere (p. 238), as the coriolis force (p. 60) bends the path of the air to the right. A wave depression of the kind considered here is also called a frontal depression, a cyclone or a low. The name cyclone is more usually used for a tropical (p. 241) hurricane (p. 70) or typhoon (p. 70).

cyclogenesis (*n*) the formation of depressions (↑).

depression
the relationship between surface air flow (black isobars) and upper air flow (500 mb surface in red contours)

the fronts of a depression

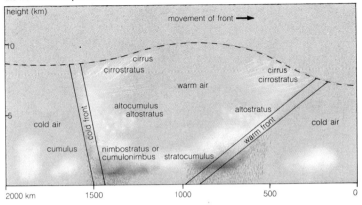

note the difference in scale between the height and distance, which makes the fronts appear much steeper than they really are

trough (*n*) a long, narrow area of low pressure. In a wave system (p. 217), a trough is an area where the curvature is cyclonic (↓).

ridge[2] a long, narrow area of high pressure. In a wave system (p. 217), a ridge is an area where the curvature is anticyclonic (↓)

cyclone (*n*) *see* depression (↑) and hurricane (p. 70).

cyclonic (adj) of a body of air flowing around a centre of low pressure, anticlockwise in the northern hemisphere (p. 238) and clockwise in the southern hemisphere.

anticyclone (*n*) an area of high pressure with light winds or calms and generally quiet weather (↓). The pressure gradient (p. 59) is outwards from the centre, so that the gradient wind (p. 61) blows clockwise in the northern hemisphere (p. 238), as the coriolis force (p. 60) bends the path of the air to the right. Also known as **high. anticyclonic** (*adj*).

weather (*n*) the state of the atmosphere, particularly the various conditions that arise, over a short time.

lee depression (*n*) an area of low pressure which forms when air is forced to cross a range of mountains or hills. On descending the range the volume of the air increases, and pressure falls. This causes convergence (p. 65) and cyclonic (p. 67) curvature of the air flow down-wind of the range.

thermal low (*n*) a low pressure system (p. 217) due to high temperatures caused by strong day-time heating of continental areas.

polar air depression a relatively small (a few hundred kilometres across), low-level low pressure system (p. 217), of either a closed, cyclonic (p. 67) circulation or one or more troughs (p. 67) within a polar (p. 239) air stream.

cold low a centre of low pressure wholly within a body of cold air, largely characteristic of the middle troposphere (p. 51). A cold low may result from occlusion (p. 65) or the cutting off of polar (p. 239) air from the main flow of cold air to the north at times when there is a low-index (*see* zonal index cycle (p. 63)) circulation in the upper westerlies (p. 63). A cut-off mass of air of this type is sometimes called a cut-off low.

cold pool = cold low (↑).

synoptic-scale (adj) of a large weather (p. 67) system (p. 217) with a life of several days, which can easily be seen on a daily weather map, e.g. a depression (p. 66) of 1500–2000 km across.

meso-scale (adj) of weather (p. 67) systems (p. 217) that are of a size in between synoptic-scale (↑) systems and, say, a single cumulonimbus (p. 57) convection (p. 52) system, e.g. a squall line (↓).

squall line a narrow band of thunderstorms (↓) sometimes several hundred kilometres long, which often forms as a result of instability (p. 55) ahead of a kata (p. 65) cold front (p. 64). It is marked by a sharp veer (p. 61) in wind direction and very gusty (p. 61) conditions.

thunderstorm (*n*) a rainstorm with lightning (↓), thunder (↓), strong wind gusts (p. 61) and perhaps hail (*see* precipitation (p. 55)). It results from strong convection (p. 52) in one or more cumulonimbus (p. 57) clouds.

thunderstorm

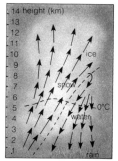

tornado

lightning (*n*) a sudden electrical flash either within a cloud or from a cloud to the Earth. Almost always the cloud form is cumulonimbus (p. 57).

thunder (*n*) the sound caused by the explosion of air that results from rapid heating by lighting (↑).

tornado (*n*) a violent, twisting windstorm, a few hundred metres across, with a tower of clouds at its centre, that passes up into a cumulonimbus (p. 57) cloud.

weather forecasting the determination of future weather (p. 67). *Short-range forecasting* is for one or two days ahead; *medium-range* or *extended* is for five to seven days ahead; and *long-range* is for a month or season ahead.

synoptic forecasting weather forecasting (↑) using synoptic-scale (↑) weather maps of surface and upper air conditions and data (p. 224) on the adiabatic (p. 54) characteristics of the atmosphere over the area concerned.

numerical forecasting weather forecasting (↑) from basic atmospheric characteristics, using computers and baroclinic (p. 64) models (p. 223) of the atmosphere. Convergence (p. 65), divergence (p. 65) and pressure distribution are calculated in 1 hour time-steps, and from the data (p. 224) produced the future weather (p. 67) pattern is worked out.

dynamical weather forecasting = numerical forecasting (↑).

statistical forecasting long-range forecasting (↑) using statistical methods. One method takes into account the effect of surface conditions, e.g. ocean temperatures, on the circulation of the upper westerlies (p. 63). A newer method examines past data (p. 224) to obtain the important climatic (p. 71) relationships, and then uses present data to determine the future weather (p. 67) in the light of those relationships.

analogue forecasting long-range forecasting (↑) using searches of past weather (p. 67) to find conditions similar to the present, i.e. an analogue or match of the present weather. Future weather should then broadly repeat that which followed the analogue, but there are many reasons why this often does not happen.

hurricane (*n*) a very strong tropical (p. 241)
cyclonic (p. 67) system (p. 217), usually about
650 km across, with pressures in the 'eye' or
centre perhaps as low as 920 mb. There is
much cumulonimbus (p. 57) cloud, up to
12,000 m, and wind speeds are often greater
than 50 ms^{-1}. Large ocean surfaces with
temperatures above 27°C are needed for
hurricane formation, which is usually in the
western parts of the Atlantic and Indian Oceans,
about 5–10° of latitude (p. 238) poleward of the
equatorial trough (↓). Convection (p. 52),
condensation (*see* condensation nucleii (p. 54)),
and an anticyclone (p. 67) in the upper
troposphere (p. 51), which allows high-level
outflow of air, are important in hurricane growth.

hurricane
plan of clouds

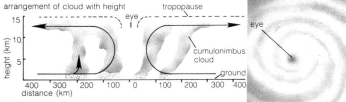

arrangement of cloud with height

typhoon (*n*) a hurricane (↑) in the western
Pacific.

easterly wave a wave in the trade wind (p. 59)
belt, travelling east to west, usually less
rapidly than the winds. Its form is a weak
trough – (p. 67) which characteristically
slopes eastward with height. Cumulonimbus
(p. 57) cloud and thundery (p. 69) showers
are found behind the line of the trough.

intertropical convergence zone that part of the
equatorial trough (↓) which lies over the
oceans, where convergence (p. 65) of the
trade winds (p. 59) is marked. Over the
continents the convergence is less continuous
in space and time, and so the name
intertropical confluence (ITC) is better. The
ITCZ proper is characterized by calms and
light winds. **ITCZ** (*abbr*).

doldrums (*n*) the older name for the intertropical
convergence zone (↑).

equatorial trough the belt of low pressure at or
near the equator (p. 238), the result of
convection (p. 52) due to heating by the Sun.

monsoon (*n*) a seasonal wind of the tropics
(p. 241) and subtropics. In summer a heated
land mass experiences low pressure and so
rain-bearing winds blow in off the cooler
ocean. The opposite happens in winter. This
simple picture does not fully explain some
of the important characteristics of the Asian
monsoon, and a knowledge of air flow in the
upper westerlies (p. 63) is necessary for a
fuller understanding.

heat island the air in and above a city which is
generally warmer than that of the rural (p. 162)
area outside. The higher temperatures
are largely due to the heat produced in a city
the nature of urban (p. 163) surfaces, and
the pollution (p. 197) of the urban atmospher
which lessens the loss of heat into space.

climate (*n*) the weather (p. 67) characteristics
of an area over a long time, usually
described by averages and statistical
measures of variation. **climatic** (*adj*).

climatic classification the grouping of climates
(↑) into different kinds according to various
characteristics, e.g. plant growth or vegetation
(p. 86), energy, precipitation (p. 55) and air
mass (p. 63) characteristics, human comfort.

continentality (*n*) a measure of the effect that
land has on air mass (p. 63) characteristics
and climate (↑). Land heats up and cools
down more quickly than the sea and the
temperature range is also much greater Thus
continental air masses have a greater
temperature range than those which form over
the sea. *See* oceanicity (↓).

oceanicity (*n*) a measure of the effect that sea
has on air mass (p. 63) characteristics and
climate (↑). Sea water does not warm up or
cool down as quickly as land, nor is the
temperature range so large. Thus air masses
which form over the sea tend to be cooler in
summer and warmer in winter than those
which form over land. *See* continentality (↑).

heat island

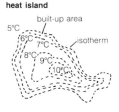

ecology (*n*) the study of how organisms (↓)
relate to each other and to the abiotic (p. 82)
part of the environment (p. 82).
ecological (*adj*).

ecosystem (*n*) a community (↓), and the abiotic
(p. 82) environment (p. 82) which supports
the community, working as a whole; i.e. an
ecological (↑) system (p. 217).

biosphere (*n*) that part of the Earth which can
support life.

biogeography (*n*) a branch of geography (p. 8)
that deals with the distribution and ecology (↑)
of plants and animals. **biogeographical** (*adj*).

zoogeography (*n*) a branch of biogeography
(↑) concerned with animals.

phytogeography (*n*) a branch of biogeography
(↑) concerned with plants.

island biogeography the ecological (↑) and
evolutionary (p. 83) biogeography (↑) of
island biotas (↓). Special attention is given to
evolutionary processes and to the way in
which species (p. 81) numbers relate to area,
i.e. to island size.

biogeographical province an area of land or
sea with a particular group of taxa (p. 77) and
communities (↓) in common. Where only
animal taxa are concerned, the name faunal
(↓) province is used, and similarly floral (↓)
provinces are recognized for plants. A *faunal
province* is also known as a zoogeographical
(↑) region. A *floral province* is also known as
a phytogeographical (↑) region. Also known
as **biogeographical realm**.

ecosystem

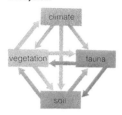

1 Arctic and sub-Arctic
2 Euro-Siberian
   A Europe **B** Asia
3 Sino-Japanese
4 W. and C. Asiatic
5 Mediterranean
6 Macronesian
7 Atlantic North American
   A Northern **B** Southern
8 Pacific North American
9 African-Indian Desert
10 Sudanese Park Steppe
11 NE African Highland
12 W. African rain forest
13 E. African Steppe
14 South African
15 Madagascar
16 Ascension and St Helena
17 Indian
18 Continental SE Asiatic
19 Malaysian
20 Hawaiian
21 New Caledonia
22 Melanesia and Micronesia
23 Polynesia
24 Caribbean
25 Venezuela and Guiana
26 Amazon
27 South Brazilian
28 Andean
29 Pampas
30 Juan Fernandez
31 Cape
32 N. and E. Australian
33 SW Australian
34 C. Australian
35 New Zealand
36 Patagonian
37 S. Temp. Oceanic islands

biogeographical province

floral provinces
Arctic Circle
Tropic of Cancer
equator
Tropic of Capricorn
Antarctic Circle

biogeographical province
faunal provinces

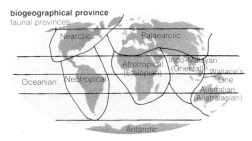

biota (*n*) the plants and animals of a given area, e.g. African biota or island biota. **biotas** (*pl*).

flora (*n*) the plant species (p. 81) which make up the vegetation (p. 86) of an area, or which relate to a particular time in the Earth's history. **floras** (*pl*).

fauna (*n*) the animal species (p. 81) of an area. **faunas** (*pl*).

organism (*n*) a living thing.

ecosystem structure the producer (p. 74) – consumer (p. 74) – decomposer (p. 74) arrangement of an ecosystem (↑). The producers or autotrophs (p. 77) form the first trophic level (p. 74) T1. The second trophic level T2 consists mainly of herbivores (p. 78). T3 and T4 are accounted for largely by carnivores (p. 78), while decomposers make up T5. Energy flows through and nutrients (p. 76) move around the various trophic levels.

ecosystem structure producer – consumer – decomposer arrangement

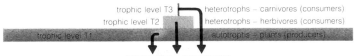

trophic level T3 — heterotrophs – carnivores (consumers)
trophic level T2 — heterotrophs – herbivores (consumers)
trophic level T1 — autotrophs – plants (producers)

heterotrophs – decomposers

biomass (*n*) the weight of live matter, usually stated as the dry matter content, in a particular area, e.g. in kg m^{-2}.

community (*n*) (1) a group of organisms (↑) that share the same environment (p. 82). (2) the plants and animals which make up the living part of an ecosystem (↑).

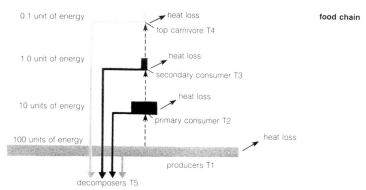

food chain

food chain the flow of energy and nutrients
(p. 76) through a chain of organisms (p. 73),
at various trophic levels (↓). Where the flow
starts from live plants, i.e. producers (↓), the
arrangement is called a *grazing food chain*;
while a *detrital food chain* begins with dead
plant and/or animal matter.

food web an arrangement of several food chains
(↑) into one unified system (p. 217).

trophic level a grouping of organisms (p. 73) in
an ecosystem (p. 72) according to similar
food needs. Thus producers (↓) belong to the
first trophic level (T1), and top carnivores to a
higher one (T4). T5 is made up of
decomposers (↓) or saprobes (↓).

niche (*n*) the part of an ecosystem (p. 72) in
which an organism (p. 73) lives, or the way of
life it follows.

producer (*n*) an autotrophic (p. 77) organism
(p. 73), which belongs to the first trophic level
(↑) in a food chain (↑).

consumer (*n*) a heterotrophic (p. 77) organism
(p. 73). *Primary consumers*, i.e. herbivores
(p. 78) eat plants; *secondary consumers*, i.e.
carnivores (p. 78), eat other animals.

decomposer (*n*) a consumer (↑) of dead organisms
(p. 73), i.e. a detritivore (p. 78). Bacteria (p. 77)
are the most important decomposers, but all
kinds play a part in biogeochemical cycles (p. 76).

saprobe (*n*) = decomposer (↑).

. **food web**

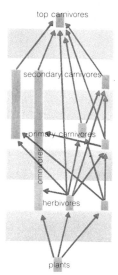

sere plagiosere

woodland

scrub

grasses

bare field

time

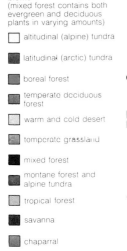

biome
(mixed forest contains both
evergreen and deciduous
plants in varying amounts)

☐ altitudinal (alpine) tundra

■ latitudinal (arctic) tundra

■ boreal forest

■ temperate deciduous
forest

☐ warm and cold desert

▨ temperate grassland

■ mixed forest

■ montane forest and
alpine tundra

☐ tropical forest

■ savanna

■ chaparral

sere (*n*) a particular kind of succession (p. 76).
Each community (p. 73) in a sere is called a
seral stage. A *prisere* is a succession from a
new land surface, e.g. *psammosere*, which
starts on sand, and *halosere*, which begins in
salt-rich environments (p. 82). A *plagiosere*
is a sere shaped by human action. **seral** (*adj*).

climax (*n*) the end point of succession (p. 76),
when the community (p. 73) is in equilibrium
(p. 220) with the abiotic (p. 82) environment (p. 82)

plagioclimax (*n*) a climax (↑) shaped by humans.

biome (*n*) an ecosystem (p. 72) that covers a
large area of the Earth's surface – land or water,
e.g. savanna (p. 87) and boreal forest (p. 86).
A biome may also be described as a
formation-type (p. 86) together with the other
organisms (p. 73) that live in the formation-type.

dominant (adj) of a plant species (p. 81) which,
because of size, particularly height, controls
the environmental (p. 82) conditions for other
plant species in the same community (p. 73).
dominant (*n*).

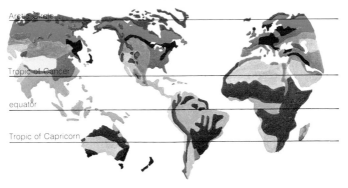

Arctic Circle

Tropic of Cancer

equator

Tropic of Capricorn

ecosystem dynamics the processes at work in
an ecosystem (p. 72), particularly to do with
food chains (p. 74), biogeochemical cycles
(↓), succession (↓), feedback (p. 220) and
homeostasis.

biogeochemical cycle the circular path followed
by a non-renewable resource (p. 191) through
the living and non-living parts of an ecosystem
(p. 72). Also known as **nutrient cycle**.

biogeochemical cycle
phosphorus

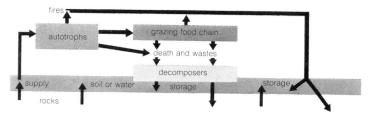

nutrient (*n*) a food substance needed by plants
and animals.

biological productivity a measure of the amount
of energy or organic matter fixed by one or
more organisms (p. 73) over a given length of
time. *Primary productivity* is that accounted for
by plants, and may be further sub-divided into
gross primary productivity and *net primary
productivity*. The former relates to all the
energy fixed by the plant, while the net figure
is that which is left and stored as living matter,
after the energy to drive the plant's organic
processes is used up. *Secondary productivity*
is the energy fixed by the primary consumers
(p. 74), i.e. the herbivores (p. 74).

competition (*n*) the struggle for resources
(p. 191) between organisms (p. 73) of the
same or different taxa (↓). **competitive** (*adj*).

succession (*n*) the steady change of species
(p. 81) in an ecosystem (p. 72), as a result of
changes in the living and abiotic (p. 82) parts
of the environment (p. 82), towards climax
(p. 75) and homeostasis.

conservation (*n*) formerly just the preservation
of plants and animals, but now taken to mean
the wise use of the Earth's resources (p. 191).

taxonomy (*n*) the science of the classification of organisms (p. 73) according to structural characters of evolutionary (p. 83) importance. **taxonomic** (*adj*). *See also* taxon (↓).

taxon (*n*) a taxonomic (↑) unit or class, e.g. species (p. 80), genus (p. 80), and so on up to kingdom (p. 79), the highest taxonomic rank. **taxa** (*pl*).

Angiospermae (*n.pl.*) the angiosperms or flowering plants, a subdivision (p. 80) of the Spermatophyta (↓) and the highest forms of life in the plant kingdom (p. 79).

Gymnospermae (*n.pl.*) the gymnosperms, a subdivision (p. 80) of the Spermatophyta (↓). Examples include conifers and cycads.

Spermatophyta (*n.pl.*) the seed plants, a division (p. 80) of the plant kingdom (p. 79), made up of two subdivisions (p. 80), the Angiospermae (↑) and Gymnospermae (↑).

Pteridophyta (*n.pl.*) the spore(↓)-bearing plants, a division (p. 80) of the plant kingdom (p. 79), which include the ferns, horsetails and club-mosses, each having the taxonomic (↑) rank of a subdivision (p. 80).

Bryophyta (*n.pl.*) spore(↓)-bearing plants which, unlike the Pteridophyta (↑), lack stems. This division (p. 80) contains two classes (p. 80), i.e. liverworts and mosses.

spore (*n*) a microscopic body produced by some plants and fungi from which new organisms (p. 73) grow.

algae (*n.pl.*) the simplest kinds of plants.

bacteria (*n.pl.*) the simplest organisms (p. 73). Some are autotrophs (↓), but most are heterotrophs (p. 78). This last group are the key decomposers (p. 74) in ecosystems (p. 72). *See* Monera (p. 81). **bacterium** (*sing*).

life form the characteristic form of a fully grown organism (p. 73), usually a plant.

autotroph (*n*) an organism (p. 73) that makes food from simple, non-organic substances, using energy from the Sun. Plants are the most important autotrophs on land and thus provide the starting point for the various kinds of terrestrial (p. 238) food chains (p. 74). **autotrophic** (*adj*).

heterotroph (*n*) an organism (p. 73) which cannot make the food it needs from simple, non-organic substances. Instead heterotrophs depend, directly in the case of herbivores (↓), or indirectly in the case of carnivores (↓), on autotrophs (p. 77) for their food supply. **heterotrophic** (*adj*).

photosynthesis (*n*) the process by which plants make food using the energy of sunlight.

carnivore (*n*) a consumer (p. 74) of animals, i.e. a secondary consumer. **carnivorous** (*adj*).

herbivore (*n*) a consumer (p. 74) of plants, i.e. a primary consumer. **herbivorous** (*adj*).

omnivore (*n*) a consumer (p. 74) of plants and animals. **omnivorous** (*adj*).

detritivore (*n*) a consumer (p. 74) of dead organisms (p. 73). **detritivorous** (*adj*).

calcicole (*n*) a plant that grows well only on lime-rich soils, e.g. rendzina (p. 92). **calcicolous** (*adj*).

calcifuge (*n*) a plant that grows well only on acid soil, i.e. a pedalfer (p. 89). **calcifuge** (*adj*).

acidophyllous (*adj*) of plants which grow on acid, nutrient (p. 76)-poor soils. They have leaves with little mineral matter.

sclerophyllous (*adj*) of plants with leaves that have much cutin covering them. The substance cutin helps plants to control water loss. Generally, sclerophyllous plants have small, hard, rather thick and leathery leaves.

epiphyte (*n*) a plant which grows on another plant, but which takes no food or water from it, e.g. ivy. **epiphytic** (*adj*).

xerophyte (*n*) a plant able to grow in dry conditions, e.g. cacti. **xerophytic** (*adj*).

mesophyte (*n*) a plant with average water needs, e.g. the oak of temperate deciduous forest (p. 87), **mesophytic** (*adj*).

saprophyte (*n*) a plant which obtains food from dead or decaying matter; an example of a saprobe (p. 74). **saprophytic** (*adj*).

annual (*adj*) of a plant which grows, produces seed and then dies within one season (year).

biennial (*adj*) of a plant which grows, produces seed and then dies within two years.

deciduous (*n*) the loss of leaves from plants for parts of the year.

kingdom
the five kingdoms of life

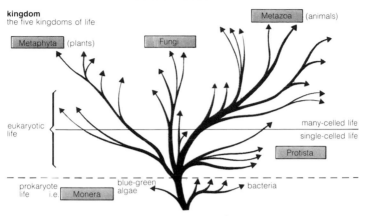

kingdom (*n*) a taxon (p. 77) of the highest taxonomic (p. 77) rank. Newer taxonomies recognize the following kingdoms: Monera (p. 81), Protista (p. 81), Metaphyta (p. 81), Metazoa (p. 81) and Fungi (p. 81).

kingdom taxonomic groups

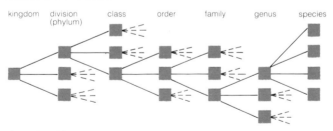

the number of lower groups which make up a higher group varies greatly

phylum (*n*) the highest taxonomic (p. 77) rank in a kingdom (↑). All the members have the same basic evolutionary (p. 83) form. In some taxonomies phyla are grouped into super phyla. A phylum is made up of classes (p. 80). *See also* division (p. 80). The Arthropoda are an example of a phylum. **phyla** (*pl*).

division (*n*) the highest taxonomic (p. 77) rank in the kingdom (p. 79) Metaphyta (↓); equal to the taxon (p. 77) super phylum (p. 79) in the kingdom Metazoa (↓). A division is made up of sub-divisions (↓). The Bryophyta (p. 77) are an example of a division.

sub-division (*n*) the highest taxonomic (p. 77) rank in a division (↑); equal to the taxon (p. 77) phylum (p. 79) in the kingdom (p. 79) Metazoa (↓). A sub-division consists of classes (↓). The Angiospermae (p. 77) are an example of a sub-division.

class (*n*) the highest taxonomic (p. 77) rank in a phylum (p. 79) or sub-division (↑). A class consists of orders (↓). The Insecta and Mammalia are examples of classes.

order (*n*) the highest taxonomic (p. 77) rank in a class (↑). An order consists of families (↓), all of which share an important, specialized evolutionary (p. 83) characteristic. Thus in the order Rodentia – the rodents – all the members have a particular kind of teeth, regardless of body shape and size.

family[1] (*n*) the highest taxonomic (p. 77) rank in an order (↑). Usually the members of a family all share some recognizable characteristic or characteristics, some common 'stamp'; e.g. the various members of the dog family – the Canidae – are all unmistakably dog-like. Families consist of genera (↓).

genus (*n*) the highest taxonomic (p. 77) rank in a family (↑). A genus consists of closely related species (↓). Thus all the various kinds of oak belong to the genus *Quercus*. The name of a genus is written in Latin, and the first letter is always a capital, e.g. *Homo*. **genera** (*pl*), **generic** (*adj*).

species taxonomic examples

kingdom	Metazoa	Metaphyta
phylum/division	Chordata	Spermatophyta
sub-division		Angiospermae
class	Mammalia	Magnoliopsida
order	Primates	Rosales
family	Hominidae	Rosaceae
genus	*Homo*	*Rosa*
species	*Homo sapiens* a human	*Rosa canina* Dog rose

species (*n*) a group of organisms (p. 73) that look very much the same as each other and which can inter-breed (reproduce sexually). Members of one species cannot inter-breed freely, if at all, with members of different species, even when in the same genus (↑) and closely related. Species names are written in Latin, after that of the genus, e.g. *Homo sapiens*. Unlike the generic name, that of the species begins with a small letter. **specific** (*adj*).

monera (*n.pl.*) the kingdom (p. 79) of prokaryotic (↓) life, consisting of two phyla (p. 79), the bacteria (p. 77) and autotrophic (p. 77) blue-green algae (p. 77) (or cyanobacteria). **moneran** (*sing*).

Protista (*n.pl.*) the kingdom (p. 79) of unicellular autotrophic (p. 77) and heterotrophic (p. 78) eukaryotes (↓), consisting of various classes (↑) and orders (↑), but no clearly separated phyla (p. 79). **protistan** (*sing*).

Metaphyta (*n.pl.*) the kingdom (p. 79) of multicellular, autotrophic (p. 77) eukaryotes (↓), namely plants. **metaphyte** (*sing*).

Metazoa (*n.pl.*) the kingdom (p. 79) of multicellular, heterotrophic (p. 78) eukaryotes (↓), namely animals. **metazoan** (*sing*).

Fungi (*n.pl.*) a kingdom (p. 79) of multicellular eukaryotes (↓), namely fungi. They are heterotrophic (p. 78) like animals, yet in other respects are more like plants. **fungus** (*sing*).

eukaryote (*n*) a unicellular or multicellular organism (p. 73), with many evolutionary (p. 83) characteristics that are absent in the more simple, i.e. prokaryotic (↓), forms of life. **eukaryotic** (*adj*).

prokaryote (*n*) the simplest kind of independent organism (p. 73), unicellular like the Protista (↑), but lacking many of the higher evolutionary (p. 83) characters possessed by these eukaryotes (↑). **prokaryotic** (*adj*).

mycorrhiza (*n*) an arrangement in which fungi (↑) enter the roots of a plant and which generally helps the growth of both organisms (p. 73). Such an arrangement is an example of symbiosis (p. 82).

symbiotic (*adj*) of two different kinds of organism (p. 73) which live together in a way which helps them both. **symbiosis** (*n*).

nekton (*n*) free-swimming organisms (p. 73) of fresh and salt water.

cryptogam (*n*) plants lacking true stems, i.e. the Bryophyta (p. 77). The Fungi (p. 81) are also included in some classifications, as are lichens, which consist of algae (p. 77) and fungi.

phanerogam (*n*) plants with true stems and roots, i.e. the Spermatophyta (p. 77) and Pteridophyta (p. 77).

environment (*n*) the conditions in which an organism (p. 73) or organisms live. The conditions include light, water, number and kind of other organisms, etc. The name is often used to describe the human habitats, where some of the conditions may be largely the result of physical geography (p. 8) or may be man-made, such as buildings, roads or even social systems (p. 217). **environmental** (*adj*).

habitat (*n*) the place or particular environment (↑) in which an organism (p. 73) lives, e.g. a woodland habitat.

limiting factor a substance or object, or the lack of it, which slows down or stops an ecological (p. 72) process, or which lessens the numbers or spread of an organism (p. 73).

pelagic (*adj*) of the open ocean.

demersal (*adj*) of organisms (p. 73), particularly fish, found near the sea floor.

benthic (*adj*) of the floor of the ocean or sea, or of a lake or river. **benthos** (*n*), life on the floor of the ocean, sea, etc.

neritic (*adj*) of the oceans above the continental shelves (p. 104).

aerobic (*adj*) of conditions or organisms (p. 73) in which free oxygen is present.

anaerobic (*adj*) of conditions or organisms (p. 73) in which free oxygen is not present.

free oxygen oxygen that can react with chemical compounds (p. 10).

abiotic (*adj*) non-living; often used with respect to the environment (↑).

biotic (*adj*) of life.

edaphic (*adj*) of the soil, particularly those characteristics which have an effect on plants and other organisms (p. 73).

eutrophic (*adj*) of nutrient (p. 76)-rich conditions for organic growth. *See also* eutrophication (p. 198).

oligotrophic (*adj*) of nutrient (p. 76)-poor conditions.

nature reserves protected ecosystems (p. 72), natural and man-made, where the aim usually is to effect little change.

evolution (*n*) the change in organisms (p. 73) through time, particularly as the environment (↑) changes. Such change allows organisms to remain well suited, or even to become better suited, to the environment. Organisms that are unable to match the speed of environmental change risk extinction (p. 84). **evolutionary** (*adj*).

adaptive radiation the evolution (↑) of various new taxa (p. 77), particularly in relation to life form (p. 77), from a common parent taxon to fill different niches (p. 74). Adaptive radiation can happen at one or more taxonomic (p. 77) levels, from species (p. 81) to kingdom (p. 79).

adaptive radiation
mammals

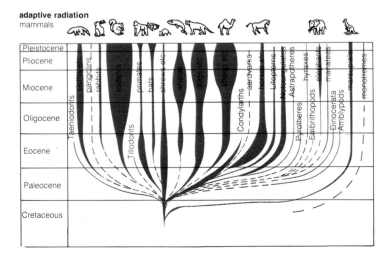

natural selection an evolutionary (p. 83) process. Because of competition, illness, etc., only certain members of a species (p. 81) are able to produce young. Thus the characteristics that permitted these members to succeed are passed on to the next ones, and so help to shape the evolution of the species. In other words, the loss or continuation of characteristics in a species is determined by natural selection.

speciation (*n*) the evolution (p. 83) of new species (p. 81). Where the new species arise as a result of the geographical break-up of the parent species, the speciation is geographical or allopatric. It is uncertain how far new species can arise in the absence of such geographical separation, i.e. by sympatric speciation.

co-evolution (*n*) the joint evolution (p. 83) of unrelated taxa (p. 77) which results in the increased competitive (p. 76) abilities or 'fitness' of the organisms (p. 72) concerned. For example, insects and flowering plants have had a great evolutionary effect on each other.

domestication (*n*) the controlled evolution (p. 83) of plants and animals by man for special purposes, e.g. increased size of fruit or grain.

extinction (*n*) the complete disappearance or loss of a taxon (p. 77) from the Earth.

dispersal route the track or path taken by an organism (p. 73) or group of organisms in spreading to new areas. The track may be over both land and water.

corridor dispersal route a dispersal route (↑) which allows free movement of many different taxa (p. 77) in both directions.

filter dispersal route a dispersal route (↑) which allows the movement of a particular group of taxa (p. 77) only, because of some limiting factor (p. 82), e.g. cold, lack of water.

sweepstake dispersal route a chance dispersal route (↑) along which very few taxa (p. 77) move, usually well separated in time.

cosmopolitan (*adj*) of taxa (p. 77) with a more or less world-wide distribution.

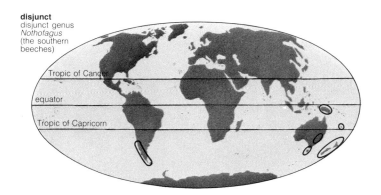

disjunct
disjunct genus
Nothofagus
(the southern
beeches)

Tropic of Cancer

equator

Tropic of Capricorn

disjunct (*adj*) of a taxon (p. 77) living in two or
more clearly separated areas, where the
distance between each area is much greater
than can be accounted for by natural spread
under present conditions. **disjunction** (*n*).

endemic[1] (*adj*) of a taxon (p. 77) that lives in a
particular area. Generally the word is used
when the area is unusually small in relation to
the taxonomic (p. 77) rank concerned.

Wallace's Line a line (imaginary) that separates
two faunal provinces (p. 72), namely the Indo-
Malayan and Australian (or Notogaean).

holarctica (*n*) a biogeographical province
(p. 72) containing the northern temperate
(p. 241), boreal (p. 86) and arctic (p. 239)
biota (p. 73). **holarctic** (*adj*).

Palaearctic (*n*) the Eurasian faunal province (p. 72).

Nearctic (*n*) the North American faunal province
(p. 72).

Neotropical (*adj*) of the South American faunal
province (p. 72).

Afrotropical (*adj*) of the faunal province (p. 72)
of Africa, south of the Sahara.

Beringia (*n*) the name given in biogeography
(p. 72) to the Bering Straits and the land around
them, which was an important dispersal route (↑).
The mammals have used Beringia, first as a
corridor dispersal route (↑) and later, when
the area cooled, as a filter dispersal route (↑).

land bridge a land mass that unites two or more
other, usually larger, areas of land. Possible
former land bridges have been claimed to
explain the disjunction (p. 85) of various taxa
(p. 77), particularly before continental drift
(see Pangea (p. 16)) was widely accepted.

vegetation (n) the plants growing together in
groups or communities (p. 73).

association (n) the part of a plant formation (↓)
where one characteristic species (p. 81) is
dominant (p. 75).

consociation (n) a climatic (p. 71) climax (p. 75)
community (p. 73) with a single dominant
(p. 75) species (p. 81).

ecotone (n) the place where two plant
communities (p. 73) mix and there is
competition (p. 76) between them.

formation
climatic climax communities

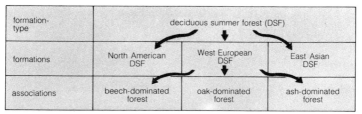

formation-type	deciduous summer forest (DSF)		
formations	North American DSF	West European DSF	East Asian DSF
associations	beech-dominated forest	oak-dominated forest	ash-dominated forest

formation (n) a clear, geographical part or
division of a formation-type (↓).

formation-type (n) a large area of world
vegetation (↑) with the same dominant (p. 75)
life form (p. 77) throughout, e.g. tundra (↓) or
rainforest (↓). See also biome (p. 75).

synusia (n) a set of plants with the same life form
(p. 77), each filling much the same ecological
(p. 72) niche (p. 74).

boreal forest (n) a holarctic (p. 85) formation-
type (↑) or biome (p. 75) with coniferous
(gymnosperm (p. 77)) trees dominant (p. 75).
The boreal forest lies immediately south of
the tundra (↓).

taiga (n) = boreal forest (↑).

carr (n) a seral (p. 75) wood on damp or wet
soils usually including species (p. 81) such
as alder and willow as dominants (p. 75).

chaparral (*n*) a sclerophyllous (p. 78) formation
(↑) in California. Chaparral is plagioclimax
(p. 75) vegetation (↑) which is ecologically
(p. 72) very similar to maquis (↓).

desert (*n*) an area that receives little or no rain,
and so supports a poor vegetation (↑)-cover
– if any at all. Desert may be thought of as a
biome (p. 75) or as a formation-type (↑).

garrigue (*n*) a French name for plagioclimax
(p. 75) low scrub (↓) of Mediterranean areas,
particularly on limestone (p. 12).

maquis (*n*) a French name for plagioclimax
(p. 75) tall scrub (↓) of Mediterranean areas.

rainforest (*n*) a tropical (p. 241) biome (p. 75)
or formation-type (↑) consisting of the
American, African and Indo-Malayan formations
(↑). The biomass (p. 73) and biological
productivity (p. 76) of rainforest are greater
than those of any other ecosystem (p. 72).

savanna (*n*) a tropical (p. 241) biome (p. 75) or
formation-type (↑) with grasses dominant (p. 75).
Much appears to be climatic (p. 71) climax
(p. 75) where there is marked seasonal dryness,
but some is undoubtedly plagioclimax (p. 75).

scrub (*n*) vegetation (↑) or habitat (p. 82) in which
bushes and/or small trees are dominant (p. 75).

temperate grassland a mid-latitude (p. 238) biome
(p. 75) or formation-type (↑) in which grasses
are dominant (p. 75). Several formations (↑) are
recognized, namely North American prairie,
Eurasian steppe, South American pampas and
South African veld. Although mainly climatic (p. 71)
climax (p. 75), some large areas of temperate
grassland are thought to be plagioclimax (p. 75).

tundra (*n*) an arctic (p. 239) biome (p. 75) or
formation-type (↑) beyond the boreal forest
(↑) in Eurasia. Grass-like plants and true
grasses are dominant (p. 75) in the vegetation
(↑), and the soils are widely subjected to
gleying (p. 91) and permafrost (p. 36) activity.

temperate deciduous forest a mid-latitude
(p. 238) forest consisting mainly of trees
which lose their leaves in the winter months.

deciduous summer forest = temperate
deciduous forest (↑).

soil (*n*) material produced by weathering (p. 19) and the activities of organisms (p. 73), and which consists of both organic and inorganic substances, water and air.

zonal soils their relationship to vegetation and climate

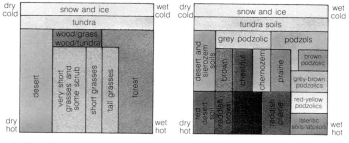

soil classification the grouping of soils according to the presence or absence of key soil horizons (p. 90) into soil types, or according to various characteristics, e.g. colour, soil reaction (p. 90) or soil texture (p. 91).

soil classification the American 'comprehensive soil classification' (or the 'seventh approximation') in relation to some zonal and azonal soils

7A	zonal/azonal soils
entisol	azonal
vertisol	swelling clay soils or grumusols
inceptisol	young brown earths and gleys
aridisol	desert and red desert soils
mollisol	grassland and forest soils with a mull humus, e.g. chernozem, chestnut, prairie
spodosol	podzols
alfisol	grassland and forest soils lacking a mull humus, e.g. grey podzolic, grey-brown podzolics
ultisol	red-yellow podzolics
oxisol	lateritic soils or latosols
histosol	peat soils

azonal (*adj*) of young soils, characterized by little-changed, i.e. fresh, parent material (p. 90). There are three groups: lithosols, formed on rock; regosols, on loose material; and alluvisols, on alluvium (p. 13).

zonal (*adj*) of the main soil types or groups, which account for large parts of the land surface, e.g. brown earth (p. 92). Generally the zonal characteristics are largely determined by the effects of climate (p. 71) and the formation-type (p. 86) in question.

generalized major soil groups of North America

☐ tundra soils

▨ peaty gleyed podzols

▨ podzols

■ grey-brown podzolic soils

▨ chernozems

▨ chestnut soils

■ complex mountain soils

☐ desert soils

☐ red and yellow podzolic (sub-tropical) soils

▨ brown earths

generalized soil groups of Britain

■ peats + peaty gleyed podzols

▨ iron podzols

▨ acid brown soils

☐ brown earths with neutral or alkaline profiles

▨ rendzinas + brown calcimorphic soils

intrazonal (*adj*) of soils that in any particular area differ from the zonal (↑) types because of special conditions, e.g. the presence of limestone as a parent material (p. 90) which gives rise to rendzinas (p. 92); or too much water in the soil profile (p. 90), which results in gleying (p. 91).

pedocal (*n*) a soil with an alkaline reaction because of the presence of free calcium carbonate, especially in the illuvial (p. 91) horizons (p. 90).

pedalfer (*n*) a soil with an acid reaction, which as a result experiences enrichment of iron and aluminium sesquioxides (p. 90) in the illuvial (p. 91) horizons (p. 90) and some breakdown of clay (*see* soil texture (p. 91)).

catena (*n*) a group of topographically (p. 22) related soils formed in the same lithological (p. 12) material. The grouping includes the soils on the upper and lower parts of a particular slope or closely related set of slopes.

catena

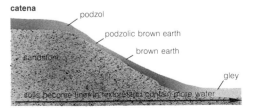

podzol

podzolic brown earth

brown earth

sandstone

gley

soils become finer in texture and contain more water

cation exchange capacity a measure of the soil's ability to store and supply cations (positive ions). The clay (p. 12) and humus (↓) content of the soil together with soil reaction (↓) determine the cation exchange capacity. **CEC** (*abbr*).

humus (*n*) the fully decayed organic material, dark in colour, on or in the soil. The common humus groups are: *mull*, which is thoroughly mixed into the 'A' horizon (*see* soil horizon (↓)); *mor*, which rests on the 'A' horizon; and *moder*, which is partly mixed into the 'A' horizon. The last two kinds are made up of: a litter layer (L) of the relatively fresh remains of plants and animals; a fermentation layer (F), where the organic matter is partly decayed; and the humus (H) proper.

horizon (*n*) a faint to clear band of soil resting on or in between other such bands. Horizons that have suffered eluviation (↓) are known as 'A' horizons, those that have experienced illuviation (↓) are named 'B' horizons, while the 'C' horizon consists of little-weathered (p. 19) parent material (↓).

laterite (*n*) an horizon (↑) of hydrated (p. 19) iron and aluminium oxides (p. 10), which may or may not be hard, often present in tropical (p. 241) red soils or red earths. Soft laterite is also known as **plinthite**.

parent material (*n*) the product of rock weathering (p. 19), in which soil forms.

sesquioxide (*n*) a chemical compound (p. 10) of oxygen (O) and a metal, with half as much oxygen again as there is metal. Thus for iron (Fe) we have Fe_2O_3, and for aluminium (Al) Al_2O_3. These two examples are the end-product of weathering (p. 19) and so are common in soils.

soil profile the arrangement of horizons (↑) from surface to rock below the parent material (↑).

soil reaction a measure of the degree to which a soil is acid, alkaline (i.e. the opposite of acid) or neutral (i.e. neither acid nor alkaline). The reaction is measured in pH units: > 7 = alkaline, < 7 = acid.

humus types

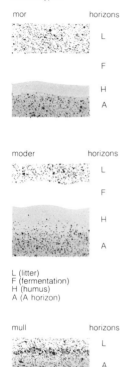

mor — horizons — L

F

H

A

moder — horizons — L

F

H

A

L (litter)
F (fermentation)
H (humus)
A (A horizon)

mull — horizons — L

A

soil structure

prismatic

columnar

angular blocky

subangular blocky

platy

granular/crumb

gley soil profile

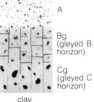

horizons

A

Bg (gleyed B horizon)

Cg (gleyed C horizon)

clay

sol lessivé/argillic brown earth soil profile

horizons

L
A
Eb (eluviation of clay)

Bt (illuviation of clay)

Cg (gleyed C horizon)

sandstone

soil texture

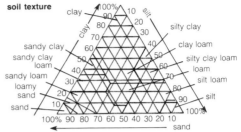

soil texture the character of a soil as determined by the amounts of clay (p. 12), silt (p. 12) and sand (p. 11) present in it.

soil structure the arrangement or grouping of soil materials into bodies called peds. The main types of ped are recognized according to shape.

calcification (*n*) the laying down of calcium-rich materials in the soil profile (↑) because of weak leaching (↓).

eluviation (*n*) the washing down of material, especially from the upper soil horizons (↑). **eluvial** (*adj*).

chelation (*n*) the union of a metal with an organic substance. Water passing through surface humus (↑) may bring about chelation and so move metals like iron down the soil profile (↑). This special kind of eluviation (↑) is called cheluviation.

ferrallitization (*n*) the enrichment of the soil in iron and aluminium oxides, due to weathering (p. 19), especially characteristic of the tropics (p. 241). **ferrallitic** (*adj*).

gleying (*n*) the changes in the iron oxides (p. 10) in soils with too much water present, resulting in the oxides becoming grey coloured rather than the usual browns, red browns or reds. Soils with these characteristics are called gleys. **gleyed** (*adj*).

illuviation (*n*) the laying down of material in the lower horizons (↑) of the soil profile (↑), e.g. in podzolization (p. 92). **illuvial** (*adj*).

leaching (*n*) the loss of soil material in water moving through the soil profile (↑).

lessivage (n) the eluviation (p. 91) and then the illuviation (p. 91) of clay. Soils that share these effects are called *sols lessivés*, lessivated brown earths (↓), or argillic brown earths.

mineralization (n) the early part of the process of decay which produces humus (p. 90).

peat (n) a partly decayed organic material, mainly from plants, formed in anaerobic (p. 82) conditions, which slow down or stop mineralization (↑). **peaty** (adj).

podzolization (n) the strong eluviation (p. 91) of sesquioxides (p. 90) and clay (p. 12), followed by illuviation (p. 91) of the sesquioxides and possibly later of humus (p. 90) as well. Soils that have experienced podzolization are called *podzols*, spodosols, etc, depending on the soil classification (p. 88). It happens under acidophyllous (p. 78) vegetation (p. 86). **podzolic** (adj).

peaty gleyed podzol a soil of temperate (p. 241) uplands with a peaty (↑) surface horizon (p. 90), resting on a grey 'A' horizon which in turn rests on a layer of iron illuviation (p. 91). The grey colour of the 'A' horizon is due to both podzolization (↑) and gleying (p. 91), caused by acidic waters from the peat.

salinization (n) the enrichment of the soil by salts of sodium, potassium and calcium, especially in dry areas. The various soils which result are: white alkalic or solonchak soils; black alkali or salonetz; and soils with characteristics in between these two, namely solodized solonetz.

brown earth the middle latitude (p. 238) forest soils coloured by iron oxides and belonging to the pedalfer (p. 89) group. Depending on soil reaction (p. 90), brown earths are subdivided into acidic and calcimorphic brown earths.

chernozem (n) a dark, deep, humus (p. 90)-rich soil of the temperate grassland (p. 87) biome (p. 75), belonging to the pedocal (p. 89) group.

rendzina (n) a thin, humus-rich intrazonal (p. 89) soil on limestone (p. 12), with an alkaline soil reaction (p. 90). Under grass, rendzinas are brown, under woodland they are much darker.

podzol soil profile

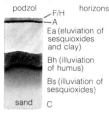

podzol   F/H   horizons
A
Ea (eluviation of sesquioxides and clay)
Bh (illuviation of humus)
Bs (illuviation of sesquioxides)
sand   C

peaty gleyed podzol soil profile

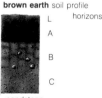

horizons
O thin peat
Eag (gleying and podzolization cause eluviation of iron)
Bfe (strong illuviation of sesquioxides)
Bs (illuviation of sesquioxides)
C
drift of sandstone waste

brown earth soil profile

L   horizons
A
B
C
sandstone

chernozem soil profile

A   horizons
Btca (illuviation of clay and calcium carbonate)
animal hole (krotovina)
loess   Cca

rendzina soil profile

A   horizons
C chalk/pure limestone

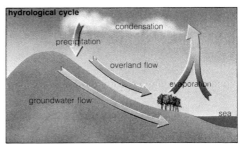

hydrology (*n*) the study of the state and behaviour of water at or near the Earth's land surface. **hydrological** (*adj*).

hydrological cycle the movement of water between various stores (↓) on and near the Earth's surface. The movement may be examined on various scales, from that of the whole Earth down to that of the drainage basin (p. 26).

store[2] (*n*) a volume of water which hardly changes for quite a long time. It is an important part of the hydrological cycle (↑). *See also* store[1] (p. 219).

reservoir (*n*) an area where the flow of water is held back by the action of man. A reservoir may be used to store water, to control stream discharge (p. 98) or to prevent a flood (p. 99).

catchment area the area which receives rain and which therefore supplies water to a stream. Also known as **drainage basin**.

watershed (*n*) an imaginary line which follows the highest points between one drainage basin (p. 26) and the next. Known in the USA as '**divide**'.

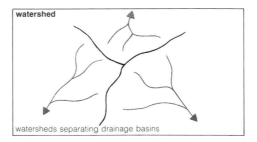

watersheds separating drainage basins

evaporation (*n*) the loss of water into the air as a
result of adding heat. This causes the water
molecules to move more quickly, and so those
near the surface may rise into the air above.
evaporate (*v*).

transpiration (*n*) the loss of water molecules
from a plant, especially from its leaves, into
the air. Together with evaporation (↑) it is an
important means of water loss in the
hydrological cycle (p. 93). **transpire** (*v*).

groundwater (*n*) the water that is found under
the surface of the Earth where all pore spaces
(p. 13) are filled. It moves slowly towards
streams or the ocean.

water-table (*n*) the surface of the zone (p. 241)
below the ground where spaces within the
rock are completely filled by water. Springs
(p. 98) are found where it meets the land
surface. It may be determined by noting the
level of water in wells (↓).

piezometric surface the artesian aquifer of the London basin. (The piezometric surface is now
much lower)

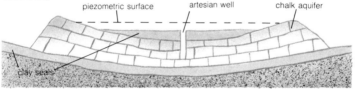

piezometric surface   artesian well   chalk aquifer

clay seals

piezometric surface the level to which water will
rise in artesian (↓) wells (↓). It is not the same
as the water-table (↑) of an unconfined
aquifer (↓).

vadose (*adj*) of a zone (p. 241) beneath the land
surface where open spaces within the rock
and soil contain air as well as water.

phreatic (*adj*) of a zone (p. 241) below the water-
table (↑) where spaces in the rock are
completely filled with water.

aquifer (*n*) a rock such as chalk (p. 17) and
sandstone (p. 11) which holds and supplies
large amounts of water. There are various
kinds: confined aquifer (↓), perched aquifer
(↓), unconfined aquifer (↓).

aquifer some types

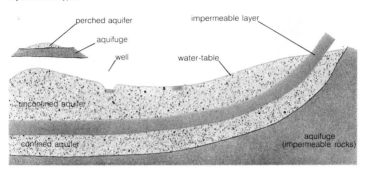

artesian (*adj*) of the condition when water is held
in a confined aquifer (↓) and so is under
pressure. Such water may escape through an
artesian well (↓).

well (*n*) a man-made hole that is dug below the
water-table (↑) to provide a store (p. 93) of
water.

perched aquifer an aquifer (↑) that is found
when a small layer of impermeable (p. 13)
rock supports a small body of groundwater (↑).

confined aquifer an aquifer (↑) that is found
when the upper surface of water is held down
by a bed of impermeable (p. 13) rock.

unconfined aquifer an aquifer (↑) that is found
when the water-table (↑) is the upper surface
of water in the rock.

aquiclude (*n*) a rock which may slowly take up
water, but which only supplies small amounts
to wells (↑) or springs (p. 98).

aquifuge (*n*) a rock which is not able to store
water because it is impermeable (p. 13).

meteoric water the water in the ground that fell
as precipitation (p. 55).

connate water the water contained in a rock and
which was formed at the same time as the rock
itself. It often contains much salt.

juvenile water the part of the water of the mantle
(p. 8) that has escaped to the ground surface.
It is then added to the world's hydrological
cycle (p. 93) as new water.

soil moisture all the water in the zone (p. 241) above the water-table (p. 94), that is the vadose (p. 94) zone.

field capacity the largest amount of water that the soil can hold.

capillary (*n*) a fine tube-like path through a rock or soil along which water may move.

runoff (*n*) the movement of water over and through hillslopes after it has fallen as rain or snow.

runoff models two important models (p. 223) which have been put forward to explain how water flows over and through hillslopes after precipitation (p. 55). (1) *Hewlett runoff model*: the sideways flow of water just below the ground surface, together with the growth of channels (p. 23), accounts for most runoff (↑). (2) *Horton runoff model*: this points to the importance of infiltration (↓) and says that where this is low, most runoff (↑) will take place over the ground surface.

interception (*n*) the ability of plants and objects, such as buildings and roads, to store precipitation (p. 55) on their surfaces for a short time.

infiltration (*n*) the movement of water into the soil. The speed of this process is controlled by, e.g. the nature of the plant cover and the degree to which the soil is permeable (p. 13).

infiltration capacity the fastest speed at which water can move into a given soil. It may be measured in centimetres per hour (cm h^{-1}).

overland flow the movement of water over the land surface towards a stream channel (p. 23). It flows either as a sheet or as a number of small threads which separate and then join.

saturated overland flow the flow that occurs over a land surface when all the open spaces in the soil and rock below are filled with water.

depression storage that amount of water which is stored in small hollows in the surface of a hillslope when rain is heavy.

quickflow (*n*) the rapid runoff (↑) which is made up of direct channel (p. 23) precipitation (p. 55), surface runoff, and rapid interflow (↓), including pipeflow (↓). It makes up the largest part of stream discharge (p. 98) during a flood (p. 99).

runoff models

heavy rain Horton overland flow

interflow (Hewlett)

interception

rainfall

interception

soil moisture

infiltration capacity
examples

soil	infiltration capacity (cm h^{-1})
clay loam	0.3 – 0.5
silt loam	0.8 – 1.5
loam	1.3 – 2.5
loamy sand	2.5 – 5.1

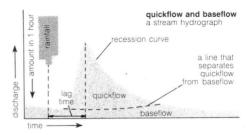

quickflow and baseflow
a stream hydrograph

recession curve

a line that separates quickflow from baseflow

rainfall

amount in 1 hour

discharge

lag time

quickflow

baseflow

time

baseflow (*n*) the amount of water that moves under the ground surface to a stream channel (p. 23). Together with overland flow (↑) it makes up all the discharge (p. 98) from a drainage basin (p. 26).

interflow (*n*) the sideways movement of water through the soil and so to a stream channel (p. 23). It takes place when the soil is less permeable (p. 13) at depth and when the ground surface slopes.

pipeflow (*n*) the movement of water through natural pipes below the ground, which measure from a few millimetres to several metres across. They may be old root holes, may be caused by animals or by water erosion (p. 20), or may simply be empty spaces that are joined together. Pipeflow is part of interflow (↑).

throughflow (*n*) the water which flows down a hillside within the soil, and then enters a stream channel (p. 23). Its speed is about $20–30$ cm h^{-1}

Darcy's law the velocity of the water (v) (i.e. flow rate divided by the area across the flow path) is equal to the product of the permeability (p. 13) of the rock (k) and the slope of the water surface in the rock (s),

$$\text{i.e.} \quad v = ks$$

This law was put forward by Henri Darcy in 1856.

percoline (*n*) a path through the soil along which water movement is relatively quick. It is often found at the head of a stream tributary (p. 27), near a stream channel (p. 23) and in a hollow.

spring

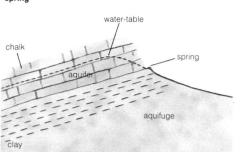

- chalk
- water-table
- aquifer
- spring
- aquifuge
- clay

spring (*n*) a place where water appears naturally at the ground surface. It is found where the water-table (p. 94) meets the land surface. A hot spring arises when water has been heated inside the Earth.

Penman formula a way of calculating the amount of evaporation (p. 94) from a surface covered by short grass.

recession curve the descending part of a graph (p. 224) relating stream discharge (↓) to time, before, during and after a storm.

stage (*n*) the height of the water flow above the floor of a river channel (p. 23). *Bank-full stage*, when the channel is completely filled with water, is important because it controls channel shape.

hydrograph (*n*) a graph (p. 224) which shows how the discharge (↓) of a stream changes.

lag time the time that passes between an input (p. 220) and the following output (p. 220) of a system (p. 217). For example, the lag time of a river valley system is the time that passes between greatest rainfall intensity and largest river discharge (↓).

discharge (*n*) the quantity of water that flows past a fixed point in a river channel (p. 23) in a certain time. It is often measured in cumecs (m^3s^{-1}). **discharge** (*v*).

regime (*n*) the changes that take place in the discharge (↑) of a stream, usually during a year. It is related closely to climate (p. 71).

regime examples

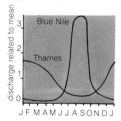

discharge related to mean

Blue Nile

Thames

J F M A M J J A S O N D J

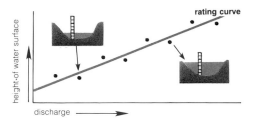

rating curve

weir
a weir set in a river channel

rating curve a graph showing how stream
discharge (↑) is related to the height of the
water surface.

weir (*n*) a plate that is placed across a stream or
flume (↓) at an angle of 90° to the direction of
flow and over which the water moves. Because
it has a regular shape it may be used for
working out discharge (↑).

flume (*n*) a man-made channel (p. 23) through
which water flows. It may be set in a stream
for working out discharge (↑), or it may be
used indoors for studying the behaviour of
flowing water (*see* model (p. 223)).

flood (*n*) (1) the largest discharge (↑) of a stream
during a year, or (2) that discharge of a stream
which is too great for the channel (p. 23) to
carry and so spreads out over the flood plain
(p. 26). **flood** (*v*).

recurrence interval the average time that
passes before an event of a given size is
repeated. The calculation is often worked out
for river floods (↑), whose recurrence interval
(in years) is arrived at by dividing the length of
study (in years) by the number of flood events.

recurrence interval discharge m³s⁻¹

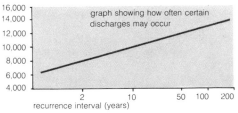

graph showing how often certain
discharges may occur

flood hydrograph a graph (p. 224) which shows how stream discharge (p. 98) changes before, during and after flood conditions.

hydraulic geometry the study of the link between stream discharge (p. 98) and stream properties such as channel (p. 23) depth, width and slope, and speed of water flow.

turbulent flow the kind of flow that shows movement at an angle to the general direction, and small changes in speed. It takes place when the Reynold's number (↓) is high, i.e. above about 750.

rapid flow turbulent flow (↑) that takes place when the Froude number (↓) is greater than one.

tranquil flow turbulent flow (↑) that takes place when the Froude number (↓) is less than one.

laminar flow the kind of flow in which the liquid, especially water, moves along paths parallel to the sides of its channel (p. 23). It happens when flow speeds are low and the channel sides are smooth.

Reynold's number a measure of the effect of viscosity (how easily a substance runs) on the flow of water. It is calculated as follows:

$$R = \frac{VD}{v}$$

where R is the Reynold's number, V is the velocity of flow, D is the water depth, and v is the viscosity. When R is low (< 500) i.e. the effect of viscosity is high, flow is laminar (↑). When R is high (> 750) i.e. the effect of viscosity is low, flow is turbulent (↑).

Froude number a measure of the effect that gravity has on the flow of water. It is calculated as follows:

$$F = \frac{V}{\sqrt{gD}}$$

where V is the velocity of water flow, g is the force of gravity, and D is the water depth. *See* rapid flow (↑), tranquil flow (↑).

Manning equation an equation which shows how the velocity of water flow is controlled by the water's depth, the slope of its bed, and the degree to which the channel (p. 23) sides and floor are smooth.

hydraulic geometry
showing how various stream properties change with discharge at a station

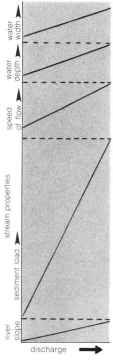

flow the nature of water flow in an open channel

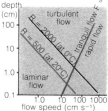

oceanography (*n*) is the study of the oceans and of oceanic processes. The science may be divided into studies of the physical, chemical, biological or geological characteristics of the oceans.

ocean (*n*) (1) a large water mass (p. 104), other than a lake or sea, with land all around it; (2) the water mass that covers all the Earth's surface not taken up by land; (3) a water mass that fills an ocean basin (↓). **oceanic** (*adj*).

ocean

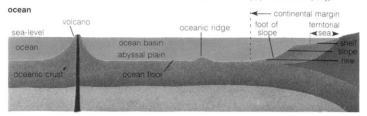

ocean floor the sea bed, including the continental shelf (p. 104) and all areas of the ocean basin (↓). *See also* oceanic crust (↓).

oceanic ridge (*n*) an under-sea mountain range found in mid-ocean and noted for a high level of earthquake (p. 208) activity and ocean floor (↑) volcanism (p. 40). Its presence supports the theory of plate tectonics (p. 15).

ocean basin (*n*) an area of ocean beyond the edge of the continental shelf (p. 104).

oceanic crust the thin outer layer of the Earth that forms the floor of the oceans. It is usually less than 8 km in thickness and, unlike the continental crust (p. 8), is continuous. *See* sima (p. 9) and sial (p. 9).

oceanic circulation the movement of water masses (p. 104) within oceans. It is closely linked to the heat of the Sun. The heat interacts with the surface of the Earth and results in winds that drive surface masses, and density changes due to cooling or evaporation (p. 94). Density changes affect the Earth's gravity field and the movement of the water masses in a deep oceanic basin (↑). This is called the thermohaline circulation system (p. 102).

thermohaline circulation system a system
(p. 217) of deep water circulation that results
from changes in temperature (*thermo*) and
salinity (p. 104) (*haline*).

thermocline (*n*) a layer of water that lies between
1 and 3 km deep, and is marked by the
maximum vertical temperature gradient (↓).

thermocline thermal stratification of the oceans

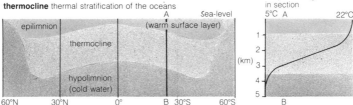

temperature gradient
in section

gyre (*n*) a large, circular pattern of oceanic
currents that move clockwise in the northern
hemisphere (p. 238) and anticlockwise in the
southern hemisphere. The pattern of gyres is
due to the effect of the prevailing westerlies
(p. 62) and the trade winds (p. 59).

gyre the surface currents
of the oceans

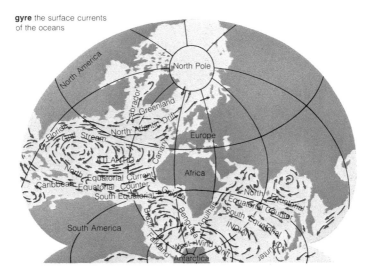

upwelling

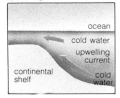

upwelling (*n*) the movement of cold water masses (p. 104) from the lower layers of the ocean towards the surface.

ooze (*n*) a deep sea deposit (p. 13) made up of the remains of microscopic organisms (p. 73) such as diatoms (siliceous (p. 9) ooze) or coccoliths (calcareous (Ca) ooze).

plankton (*n*) organisms (p. 73) that float near the ocean surface. They can be divided into phytoplankton (↓) and zooplankton (↓). Many of these organisms are microscopic but a plankton community (p. 73) may include the egg and larval forms of larger creatures.

zooplankton (*n*) very small animals, including heterotrophic (p. 78) Protista (p. 81), which float near the surface of the seas and oceans. They are linked with currents and include various unicellular animals as well as the egg and larval forms of larger animals, e.g. crustaceans and fish.

phytoplankton (*n*) very small plants, autotrophic (p. 77) Protista (p. 81), which float near the surface of seas and oceans. They are an important source of food for many other organisms (p. 73). They often form dense blooms (↓) and are linked with cold upwelling (↑) currents.

bloom (*n*) a dense layer of plants and/or animals (plankton (↑)) in the upper levels of the sea. The presence of such great numbers of these organisms (p. 73) may be linked with the upwelling (↑) of cold waters rich in food materials.

El Niño (*n*) a warm ocean current that flows along the coast of Peru every seven or fourteen years, when it replaces the colder Humboldt (Peru) current. It results in a 10°C increase in surface-water temperatures and in a decrease in plankton (↑). This in turn has a considerable effect on the fish population (p. 146) and fishing industry.

krill (*n*) tiny animals found in great numbers in the cold Antarctic waters. They are an important part of food webs (p. 74) in these waters.

continental shelf the underwater continuation
of the continents. The shelf normally stretches
200 miles in a seaward direction and its outer
boundary (p. 173) is marked by the
continental break or slope at an average depth
of 200 metres. The topography (p. 22) and
geology of the shelf are continuations of those
of the land mass it borders.

water mass a mixture of several water types (↓)
that fall within a given range of salinities (↓) and
temperatures. Salinity and temperature
together determine the density of the water
mass. The densest waters are found at the
bottom of the oceans.

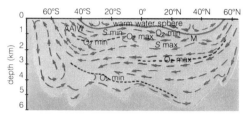

water mass major water
masses of Atlantic Ocean
identified on basis of
salinity-temperature plots

- → main direction of
  water flow
- O_2 min – minimum oxygen
  concentration
- O_2 max– maximum oxygen
  concentration
- AAIW — Antarctic
  Intermediate Water
- M —— Mediterranean water
  (flowing from east
  to west)
- S min — salinity
  minimum
- S max — salinity
  maximum

water type a body of water with a certain
temperature and salinity (↓). Freezing,
evaporation (p. 94) or heavy rainfall may each
alone or all together determine the nature of
the water type. Several water types mix to form
a water mass (↑).

salinity (*n*) a property of sea water; the degree
to which salt is dissolved in the water.

tides (*n.pl*) events linked with the effects of the
gravitational forces of the Moon and, to a
lesser degree, of the Sun, on the Earth's
surface. When the two pull in opposition they
produce low *neap tides* and when they pull
together they produce high *spring tides*.

wave
wave and variation of wave
motion at depth

variation of
motion at
depth

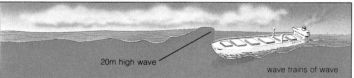

20m high wave

wave trains of wave

waves (*n.pl*) movements in the surface of water
that are linked directly or indirectly with the wind.
Oceanic waves may travel far beyond the area
where they are produced as swell; waves also
form within smaller water masses (↑). Tsunami
or seismic (p. 208) waves are the result of
under-sea earthquakes (p. 208) and often have
catastrophic (p. 208) effects on coastal areas.
Destructive waves remove more material
seawards than towards the land; *constructive
waves* deposit (p. 13) sediments (p. 11) on
beaches (p. 46). The frequency (p. 225) of
destructive waves, 13 to 15 per minute, is
usually twice that of constructive waves.

contiguous zone
coastal state waters

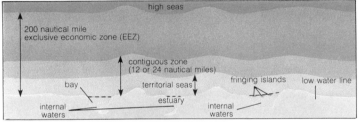

contiguous zone a protected strip of water
claimed by coastal states under the 1958
Geneva Convention. At first the zone stretched
12 miles seaward from the shoreline but this was
increased to 24 miles under the UN Convention.
exclusive economic zone a strip of water that
stretches 200 miles seawards from a country's
coastline, within which the coastal state alone
has the right to explore, exploit, conserve (p. 76)
and manage resources. The zone is drawn at
the same distance as the exclusive fishery zone
within which the state alone has the right to fish.
**United Nations Conference on the Law of the
Sea** meeting first held in 1958; the latest was
in 1982. These have established various laws
of the sea under the Geneva Conventions on
the Law of the Sea. The four conventions deal
with the Territorial sea; the High Seas; Fishing
and Conservation (p. 76); and the Continental
Shelf (↑). **UNCLOS** (*abbr*).

human geography the study of the man-made parts of the Earth's land surface and of the ways in which man is arranged as a social animal.

cultural geography the study of the ways in which people live, how these ways vary over the Earth's surface, and how the differences may be explained.

industry (*n*) the economic (p. 122) activities which produce value, usually as goods (p. 122) or services. **industrial** (*adj*).

industrialization (*n*) the growth of industries and the amount of industry in an area.

industrial location theory an attempt to explain location (↓) patterns of industrial activity. It may relate to single industries or to many. *See* Weber's location theory (↓).

location (*n*) a place where a thing is found or where an activity takes place. **locate** (*v*).

concentration (*n*) the state of a number of things or activities being gathered together in a place or area, as with industrial agglomeration (p. 108). **concentrate** (*v*).

location triangle an area used in Weber's location theory (↓) in which an industry can locate at a point where transport costs are least. Two corners of the triangle are places where raw materials are obtainable and the other corner is a market centre.

isodapane (*n*) a line which joins places with equal costs for moving goods (p. 122) or materials.

location triangle

C = market centre

M_1, M_2 = raw material points

L = location of least transport cost (point of movement minimization)

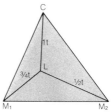

isodapane

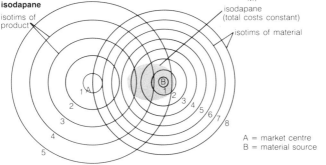

isotims of product

isodapane
(total costs constant)

isotims of material

A = market centre
B = material source

critical isodapane
labour location

P = minimum movement
point

L_1, L_2 = labour locations with
cost saving of 10

L_1 is an optimal location

L_2 is a sub-optimal location

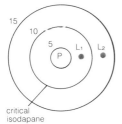

critical
isodapane

critical isodapane a line joining places where the added costs in moving away from the point of least transport costs are equalled by savings from lower-cost labour or agglomeration (p. 108).

isotim (*n*) a line which joins places with the same buying costs or distribution costs (p. 122).

Weber's location theory the model (p. 223) used by Alfred Weber (1909) to explain the pattern of industry on a large scale. It mainly uses the idea of lowest transport costs and is thus a least-cost location (↑) model. The optimum location (p. 109) is given by a location triangle (↑). Labour locations or points of agglomeration (p. 108) can also be optimum locations.

minimax location in industrial location theory (↑) the place for a firm (p. 112) where costs are lowest and where income is highest.

material index

material index (MI) = $\dfrac{\text{weight of localized raw materials}}{\text{weight of finished product}}$

MI > 1 is a pull to materials location (weight loss industry)
MI < 1 is a pull towards market location (weight gain industry)
MI = 1 location at materials, market or any point between

material index a number used in Weber's location theory (↑) which expresses the ratio (p. 224) of raw materials (p. 121) to finished products (p. 121) for manufacturing (p. 118) industry. It is a guide to the least-cost location (p. 109) and shows if an industry is pulled more to markets (p. 123) or raw materials. **MI** (*abbr*).

weight loss industry an industry where the finished product (p. 121) is very much lighter than the raw materials (p. 121), i.e. a high material index (↑), e.g. a works for making pure metals which is usually located (↑) next to raw materials to lessen transport costs.

weight gain industry an industry where the finished product (p. 121) weighs more or is bulkier (i.e. takes up more space) than the raw materials (p. 121), i.e. a low material index (↑). Examples are the brewing, baking and soft drinks industries which are usually located (↑) near the market (p. 123).

pure materials raw materials (p. 121) which lose
  little waste (p. 209) in manufacturing (p. 118)
  or processing; e.g. sand and gravel (p. 11)
  clay (p. 12) for bricks.
gross materials raw materials (p. 121) which
  lose much as waste (p. 209) during
  manufacturing (p. 118) or processing; e.g.
  minerals such as iron ore (p. 118), sugar beet.
ubiquitous materials those raw materials (p. 121)
  in Weber's location theory (p. 107) which are
  believed to be found in all places, e.g. water.
localized materials the raw materials (p. 121) in
  Weber's location theory (p. 107) which are
  found only in certain places.

deglomeration
agglomeration, deglomeration and economies of scale with size
of centre

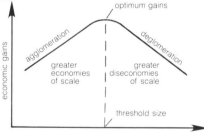

agglomeration (*n*) the concentration (p. 106) of
  industries at a place which gives economies
  of scale (p. 122) by lowering production costs
  (p. 123) or for reasons of industrial linkage
  (p. 112). **agglomerate** (*v*).
deglomeration (*n*) the movement of industries
  away from areas of agglomeration (↑) because
  of increased costs of, e.g. labour, transport
  and land. This can then give deglomeration
  economies (p. 123). **deglomerate** (*v*).
market orientation the location (p. 106) of an
  industry at or close to the market area
  (p. 123). It is usual in weight gain industries
  (p. 107) or those that need rapid response to
  market changes, e.g. the printing industry.

agglomeration
area of agglomeration at
overlap of critical
isodapanes

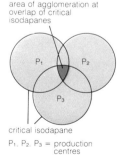

critical isodapane

P_1, P_2, P_3 = production
centres

Varignon frame

C = centre of market

M_1 = raw material 1 (of which ¾ tonne needed)

M_2 = raw material 2 (of which ½ tonne needed)

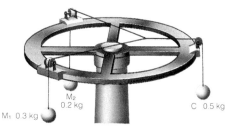

M_2 0.2 kg

C 0.5 kg

M_1 0.3 kg

Varignon frame a piece of equipment which can be used to give an answer to Weber's location (p. 107) triangle problem. The weight on the three strings is related to the pull or the amount of material required for the product.

minimum movement point the point in Weber's location theory (p. 107) where total costs of moving raw materials (p. 121) and goods (p. 122) in the manufacture (p. 118) of a product is lowest. This is the optimum location (↓) or least-cost location (↓).

optimum location the place where an industry has the best possible chance to make its costs lowest and/or its income is highest. In Weber's location theory (p. 107) this is the minimum movement point (↑).

least-cost location the place for an industry where transport costs for raw materials (p. 121) and finished goods (p. 122), and costs for other factors of production such as labour costs, are lowest. Most industrial location theories (p. 106) attempt to find this point.

profitable location a location (p. 106) where a firm (p. 112) is able to make a return on capital (p. 114). It is inside the spatial margins to profitability (p. 123).

Löschian demand cone in a model (1954) by August Lösch, a German economist, this shows how large a market area (p. 123) will be for a firm (p. 112). Demand for goods (p. 122) or services decreases with distance from the optimum location (p. 109) as transport costs rise from the place of production. Thus the price rises to a point where demand disappears.

Löschian demand cone

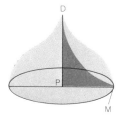

D

P

M

PD = demand or price at point of production

M = place where demand and price reach zero

DM = slope of distance decay of demand or price

locational interdependence for two suppliers and market areas

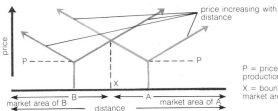

P = price at point of production

X = boundary between market areas

locational interdependence the idea in industrial location theory (p. 106) that a firm (p. 112) will choose a location (p. 106) by taking account of other firms with which it is in competition (p. 76). This is to give it a good share of the market area (p. 123) and may lead to agglomeration (p. 108) or to more dispersed patterns.

location decision the set of choices which face entrepreneurs (p. 113) in the use of resources. The largest group of decisions are concerned with an increase in the size of a plant; others are concerned with relocation, the formation of branch plants (p. 121), and the location (p. 106) of a new firm (p. 112). Location decision making is the process of assessing various locations for an industry and includes consideration of both economic (p. 122) and non-economic factors such as personal factors.

location decision

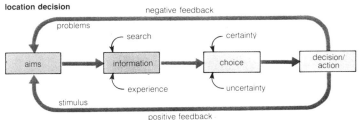

adaptive (*adj*) of the way a decision-maker behaves in an active search for the best or optimum location (p. 109), i.e. he adapts to the economic (p. 122) environment (p. 82) as seen in the behavioural matrix (↓). **adapt** (*v*).

behavioural matrix

● less good locations
● best (optimum) location

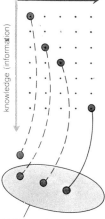

ability to use it

knowledge (information)

spatial margin
to profitibility

adoptive (*adj*) of the way a decision-maker finds a profitable location (p. 109) in the behavioural matrix (↓) by chance, i.e. is adopted by the economic (p. 122) environment (p. 82). **adopt** (*v*).

'economic man' the theoretical (p. 222) decision-maker who considers all economic (p. 122) factors based on perfect knowledge and the ability to use it in aiming for the best possible result. Also known as **'optimizer'**. *See also* 'satisficer' (↓).

'satisficer' (*n*) the decision-maker who does not attempt to gain the best possible economic (p. 122) result. The idea is used in behavioural models (p. 223) with 'satisficers' in various positions in the behavioural matrix (↓).

behavioural matrix a model (p. 223) which attempts to show the place of 'satisficers' (↑) who have various amounts of knowledge and varying ability to use it in decision-making.

isotropic plain the spatial (p. 141) surface for economic (p. 122) activity that has the same characteristics all over. Thus the relief (p. 23) and soils are the same.

environment of uncertainty the idea that decision-makers are faced with lack of complete knowledge now and in the future about factors which affect their firms (p. 112).

primary industry industry producing natural resources (p. 196) in a raw state, e.g. forestry (p. 134), agriculture and mining industries.

secondary industry industry which uses materials (raw materials (p. 121) or parts) to manufacture (p. 118) goods (p. 122). Also known as **manufacturing sector**.

tertiary industry the group of industries which offer services, e.g. finance, insurance, retail.

quarternary industry a branch of industrial activities which offers (for example) advice, training, research and development (p. 175).

research and development the activity which supports industry with new ideas for products and for better production. High technology industry (p. 113) especially uses this activity. **R & D** (*abbr*).

classification of industries the dividing of industries into groups according to kind. In Britain the Standard Industrial Classification (SIC) has twenty-seven main groups (Main Order Heading – MOH), each with a number of Minimum List Headings (MLH). A further possible classification is by broad groups; i.e (1) primary industries (p. 111), (2) secondary industries (p. 111), (3) tertiary industries (p. 111), (4) quaternary industries (p. 111)

firm (*n*) an organization which is concerned with enterprise (p. 122) and business at a place (sometimes called an establishment)

industrial market industries which buy the goods (p. 122) made by other industries

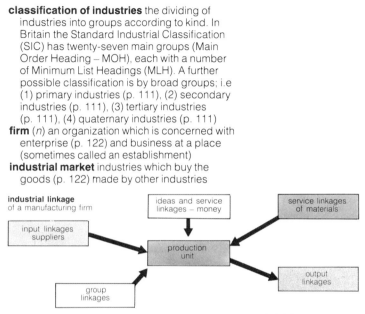

industrial linkage
of a manufacturing firm

ideas and service linkages – money

service linkages of materials

input linkages suppliers

production unit

output linkages

group linkages

industrial linkage the ties between industries which need each other's goods (p. 122) or services; e.g. component suppliers (p. 121) and the automotive industry (p. 122). It can be vertical (↓), convergent (↓) or diagonal (↓).

convergent linkage a form of industrial linkage (↑) where firms (↑) make components (p. 121) which feed separately into a main industry; e.g. car bodies, wheels, tyres, etc for the motor assembly industry.

diagonal linkage a form of industrial linkage (↑) where an industry serves many other industries; e.g. toolmakers which serve engineering industries.

vertical linkage a form of industrial linkage (↑) where, for example, industries refining metals directly serve the industries manufacturing (p. 118) metal goods (p. 122)

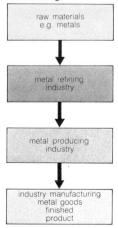

vertical linkage

raw materials
e.g. metals

metal refining industry

metal producing industry

industry manufacturing metal goods finished product

high technology industries industries which (1) develop (*see* research and development (p. 111)) or make new kinds of equipment, e.g. micro-electronic equipment such as computers, micro-chips; (2) use such equipment in production

entrepreneur (*n*) a person or group (the board of a company or corporation (p. 124)) who take the risk of starting or carrying out a business. The entrepreneur can be viewed as a decision-maker and as a factor of production. **entrepreneurial** (*adj*).

cityport industrialization the idea of industrialization (p. 106) of a port area which includes linkage (p. 219) between port activities, industrial development (p. 175) and regional planning (p. 182). The industries which are often closely linked with ports are called maritime industries

maritime industries those industrial activities which are directly or indirectly related to ports For example: (1) shipbuilding and marine engineering; (2) industries which process imported (↓) raw materials (p. 121); (3) industries which produce goods (p. 122) mainly for an export (↓) market (p. 123). Also known as **port-related industries**.

multinational companies usually the largest corporations (p. 124) whose decision-making is on a world scale and who have industrial plants (p. 115) and activities in many countries.

industrial complex analysis a way of showing linkages (p. 219) between industries in areas of concentration (p 106) of industrial linkage (↑)

infrastructure (*n*) the services and material things such as transport, water power, etc. which are needed for agricultural, industrial and other forms of economic development (p. 175). **infrastructural** (*adj*)

export (*n*) a good (p. 122) or raw material (p. 121) which is sent out from one country or region to another. **export** (*v*).

import (*n*) a good (p. 122) or raw material (p. 121) which is brought into a country or region from another. **import** (*v*).

capital (*n*) money (finance) which is a main
resource for development (p. 175). *See* capital
intensive (p. 123), venture capital (p. 123)
fixed capital (p. 123)

public sector that part of the economy (p. 122)
which is controlled by government, e.g. the coal
and rail industries in Britain

private sector that part of the economy (p. 122)
which is controlled by private enterprise (p. 122)

decentralization (*n*) the movement of industry
from an area of industrial concentration
(p. 106) to a more peripheral location (↓)
decentralize (*v*).

industrial inertia the idea that an industry can
remain in a location (p. 106) after the factors
which provided its first choice of location have
mostly disappeared; e.g. some plants of the
iron and steel industry in Britain and Pittsburgh
in the United States.

location quotient[1] a means of showing the
concentration (p. 106) of an industry in an area
such as a city or region (p. 241). A value over
1.0 means the industry is more localized in that
area than the average for all areas. **LQ** (*abbr*).

location quotient	
$LQ =$	number employed in industry (i) in area (x) as % of employment in all industries in x
	number employed in industry (i) in all areas (n) as % of employment in all industries in n
e.g. in city x	total employment = 1,000 brewing employment = 100
in region n	total employment = 10,000 brewing employment = 900
thus	$\dfrac{\dfrac{100}{1,000} \times 100}{\dfrac{900}{10,000} \times 100} = \dfrac{10}{9}$ for brewing in the city x, LQ = 1.11

footloose (*adj*) an industry which appears to have
free choice in its location (p. 106) with no special
pull to raw materials (p. 121) or markets (p. 123)
e.g. high technology industries (p. 113)

peripheral location an area for economic (p. 122)
activity which is marginal to the main centres of
agglomeration (p. 108); in Britain, e.g. Scotland,
Wales, northern England – away from the core
(p. 176) regions (p. 241) of the Southeast.

industrial park in the USA

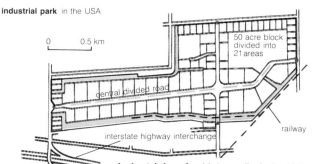

0   0.5 km

50 acre block
divided into
21 areas

central divided road

interstate highway interchange

railway

park boundary

industrial areas

surface water drainage
lines

landscaped central
road islands

railway line serving
some industrial areas

● main park road
entrances

industrial park a high-quality industrial area which
is planned with full services for a group of
industries. The idea was first used in North
America in the early 1950s. It has come to
replace names such as industrial estate (↓) in
many countries. Special kinds include
science parks (↓), air parks (an industrial area
next to an airport), distribution centres
(p. 116), energy parks (↓) and business parks
(office parks) (for non-manufacturing industries)

industrial estate a planned area complete with
services for a group of industries. The estate
is developed by an organization of either the
public or private sector (p. 161). Newer planned
areas are usually called industrial parks (↑)

trading estate = industrial estate (↑).

industrial complex a group of related members of
a given industry in an area, centred around the
main firm (p. 112) to give industrial linkage (p. 112).

science park a very high-quality planned area
where industry is related R &D (p. 111). Also
known as **research** or **technology parks**.

energy park a form of industrial park (↑) which
has a central energy source, such as a steam
or electricity generator, which is used to supply
all the firms (p. 112) in the park.

'decoplex' a name for an industrial park (↑) which
integrates (p. 124) waste recycling (p. 195) and
industry. It is short for 'development-ecological-
complex'. Also known as **'decopark'**

industrial plant a factory site with buildings used
for manufacturing (p. 118) or processing

enterprise zone an area, usually urban (p. 163),
which is set aside for industries and other
economic (p. 122) activities and where special
laws allow many restrictions to be lifted, so
that, e.g. planning is easier for industries
There are over twenty such zones in Britain;
e.g. the London Docklands Enterprise Zone.

foreign trade zone an area set aside for
economic (p. 122) activities where costs are
saved because the area is considered to be
outside the usual customs territory (where
taxes are paid) of the country concerned. This
saving attracts industries to the area, industrial
promotion. FTZs may be areas of cityport
industrialization (p. 113) or inland places

freeport = foreign trade zone (↑)

industrial corridor city of Detroit, Michigan, USA

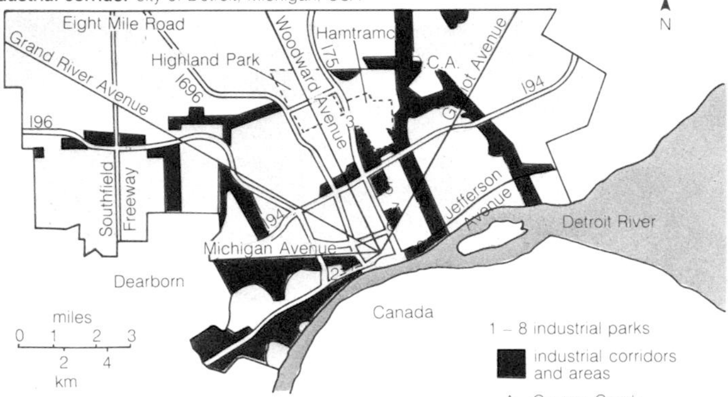

1 – 8 industrial parks

■ industrial corridors
and areas

A Conner Creek
   corridor

B Concord corridor

C Southwest corridor

D.C.A. Detroit City
   airport

I75 interstate highways

industrial corridor an area of industrial location
(p. 106) or agglomeration (p. 108) which
stretches along a line, often alongside a main
road or railway.

distribution centre a form of industrial park
(p. 115) which is especially planned for
activities that distribute products to markets
(p. 123), e.g. wholesaling, warehousing.
Such centres are usually located (p. 106)
close to main roads for fast and easy
accessibility (p. 141) to market areas (p. 123).
Also known as **distribution park**.

break of bulk a location (p. 106) where raw materials (p. 121) are processed or manufactured (p. 118) at the point of unloading from transport. This can save some of the high costs of handling raw materials by loading and unloading. Examples: iron ore (p. 118) unloaded from bulk carriers and processed at tidewater locations ($\downarrow$); a city port (p. 165) such as Rotterdam/Europoort.

tidewater location a place for location (p. 106) of economic (p. 122) activity which is on a site where there is deep water from tides (p. 104), so that large ships can be used. Such sites may be break of bulk ($\uparrow$) points and include coastal ports, estuary (p. 48) and main river locations.

greenfield site a completely new site or place for a factory or industry.

brownfield site a site for a new industrial plant (p. 115) in an area where industry has already been located (p. 106).

petrochemical complex an industrial area which is an agglomeration (p. 108) of chemical processing centred around a large oil refining (p. 122) plant. The location (p. 106) of such an industrial complex is usually a tidewater ($\uparrow$) site which is a 'break of bulk' ($\uparrow$) point for imported (p. 113) raw materials (p. 121).

'silicon valley' the name first given to the agglomeration (p. 108) of high technology industry which has grown along the corridor (p. 163) in Santa Barbara County following Route 101 south of San Francisco, California, USA. It is now used for similar areas, e.g. a part of Central Scotland is known as 'silicon glen'

'western corridor' a growing concentration (p. 106) of high technology industries along the M4 motorway in southern England. It includes towns such as Reading, Swindon and Bristol and has many R & D (p. 111) firms (p. 112).

'sunbelt' location an area of rapid growth and concentration (p. 106) of industry in a region (p. 241) of good climate (p. 71) and with a pleasant environment (p. 82). The name was first used for the growth areas in the southern states of the USA but is now used more generally.

ores (*n.pl.*) minerals which contain a desired substance in amounts worth mining. The amount of this substance in the material is expressed as grade, which may be low or high. The poorest grade that can be mined is the cut-off grade. The area of mining is known as an ore field.

manufacture (*v*) to make things in a factory. For example, a manufacturing industry is based in a factory (p. 121).

open cast mine a mine where surface earth and rock is moved to allow minerals to be dug out, e.g. many ores (↑), coal in Wyoming, USA. It is cheaper and easier than deep shaft mining.

high grading the practice in some mining of taking out the rich ores (↑) first to leave the low grade ores behind.

iron and steel industry the name for the industrial activity which joins the manufacture (↓) of iron, using ore (↑) as the raw material (p. 121), with the making of steel and steel products.

Bessemer process a way of making steel quickly in a vessel or container from liquid pig iron direct from a blast furnace (↓) with scrap (p. 120) added. It can be used to make basic steel (↓) or acid steel (↓). The process is named after Henry Bessemer (1856). More up-to-date ways are now usual; *see* basic oxygen process (↓).

pig iron the iron metal, containing small amounts of carbon and other impurities which come from a blast furnace (↓) and may be used in liquid form for steel making (↑) or cooled to form ingots (↓).

basic steel a way of making steel which uses a limestone (p. 12) lining inside the furnace (↓) to help take phosphorus out of the iron.

acid steel a way of making steel in a furnace (↓) with an inside coat of silica (p. 9) which does not take out phosphorus or sulphur from the iron. The process must therefore use haematite (an iron ore (↑)) which does not contain phosphorus.

basic oxygen process
vessel for making steel
smoke collecting hood
hole for pouring
lining
pouring position of vessel
steel shell of vessel
water-cooled oxygen input
liquid metal

blast furnace

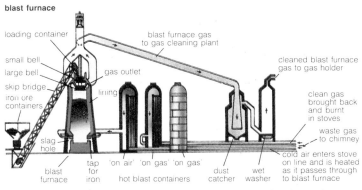

furnace a large vessel or container heated to a
high temperature, used for making iron or
steel.

blast furnace a large container which is fed with
iron ore (↑), coke (burnt coal) and limestone
(p. 12) flux (to remove impurities) into which
air is blown to take out iron oxides (p. 10).
The result is a liquid pig iron (↑) which is used
to make iron or steel.

basic oxygen process a way of making steel
from liquid iron together with scrap (p. 120).
This is done using oxygen gas blown into the
mix of raw materials (p. 121). Aluminium,
carbon and other materials can also be added.
Also known as **Linz-Donawitz (LD) process.**

ingot a large block of metal, e.g. steel or pig
iron (↑), made from the liquid metal coming
from a furnace (↑).

continuous casting a way of making steel in
which, instead of first being made into ingots
(↑), it is teemed (p. 120) directly into a
machine to make blocks or sheets of steel.

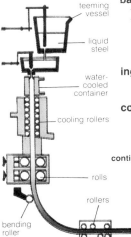

continuous casting steel

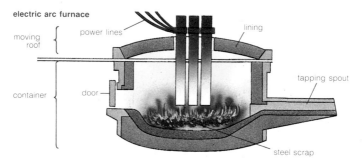

electric arc furnace

moving roof

power lines

lining

container

door

tapping spout

steel scrap

electric arc furnace a way of making steel direct from scrap (↓) in a furnace (p. 119) which uses a strong electric charge to melt the scrap. This kind of furnace is usually found in 'mini-mills' (↓).

scrap (*n*) the material used in steel-making furnaces (p. 119) which comes as a by-product from industrial processes or is other kinds of waste, e.g. old cars, ships, etc.

'mini-mill' steel mills which are not fully integrated (p. 124) plants and where steel is made from scrap (↑) in electric arc furnaces (p. 119). They feed the continuous casting (p. 119) process and light rolling mills (↓). They usually make only 500,000 to 1 million tonnes of steel each year for a regional (p. 241) market (p. 123).

teem (*v*) to pour liquid iron or steel into a container to form an ingot (p. 119).

rolling mill a special plant which makes sheets of rolled steel either from heated steel ingots (p. 119) or in the continuous casting (p. 119) process from liquid steel.

integrated plant a factory which builds nearly all the parts needed to make a finished product (↓), e.g. (1) some car-making plants which build engines and other parts to make a finished vehicle, (2) an iron and steel works which makes its own iron.

smelter (*n*) the plant used for the concentration (p. 106) and/or refining of mineral ores (p. 118), e.g. a tin or copper smelter. **smelt** (*v*).

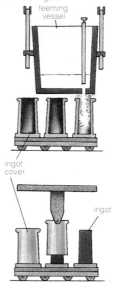

teem
teeming liquid steel to make ingots

teeming vessel

ingot cover

ingot

mineral extraction the mining of mineral ores
(p. 118). The method may be open-cast mining
(p. 118), deep mining or dredging of alluvial
(p. 13) deposits (p. 13) or re-working waste.
low grade ores rocks which contain minerals in
very low concentration (p. 106); e.g. many
valuable metal ores (p. 118) are mined when
they have less than 1% of the metal. The lowest
grade that is mined is called the cut-off grade.
beneficiation the practice of concentrating
(p. 106) low grade ores (p. 118) at a site, e.g. a
mine where they are extracted. This saves the
costs of transporting mineral waste.
assembly plant a factory which puts together
components (parts) to make a finished
product ($\downarrow$), often to serve a regional (p. 241)
market (p. 123), e.g. a car assembly plant.
branch plant a factory which is at a location (p. 106)
away from the headquarters of the firm (p. 112).
extractive industry a form of primary industry
(p. 111) producing raw materials ($\downarrow$) for processing
and manufacturing (p. 118) industries; e.g. the
mining of coal and mineral ores (p. 118).
processing industry industry which prepares raw
materials ($\downarrow$) for further manufacturing (p. 118);
e.g. food industry, mineral concentration (p. 106).
raw materials the inputs (p. 220) needed by an
industry to make a product. These may be used
first by processing industry ($\uparrow$), e.g. mineral
concentration (p. 106), or as parts used by
manufacturing (p. 118) industry.
factory the buildings which contain machinery
used by manufacturing (p. 118) industry. Also
known as **industrial plant**.
finished product the end result of a manufacturing
(p. 118) process, which may be a good (p. 122)
for the final market (p. 123) or a part for further
manufacturing or for an assembly line ($\downarrow$).
assembly line a way of manufacturing (p. 118)
goods (p. 122) in a continuous line where
parts are added at points along the moving
line. This can lower the costs of the product.
component suppliers the firms (p. 112) which
make parts for assembly industries; e.g. wheels,
tyres, etc for the automotive industry (p. 122).

automotive industry an industry which manufactures (p. 118) motor vehicles. The industry includes integrated (p. 124) plants and assembly plants (p. 121) with component suppliers (p. 121). The few companies or corporations (p. 124) in the industry are usually multinational (p. 113) and work under conditions of oligopoly (i.e. in a market where there are only a few sellers).

oil refinery the processing plant where crude oil is changed into a large number of products. Where the plant is large and many chemicals are made from these products, it is often called a petrochemical complex (p. 117).

good (n) the product of a manufacturing (p. 118) industry which is sold. See durable goods (p. 124), non-durable goods (p. 124).

economy (n) the system of production of things of value and their movement between places, so economic geography is the study of places of production and the movement of goods (↑). **economic** (adj).

enterprise (n) a business activity which results from a decision by a person or group and acts as a firm (p. 112). **enterprising** (adj).

agglomeration economies the savings made by a firm (p. 112) as a result of location (p. 106) close to other industries. The savings may be, for example, on the costs of labour, transport and buying goods (↑).

external economies the savings made by an industry by sharing services and infrastructure (p. 113) in an area of agglomeration (p. 108).

economies of scale the savings in unit costs (↓) made by an industry because of the large size of the industry or its plant. These may be from internal (↓) or external economies (↑).

diseconomies of scale the higher costs to an industry because of the over-large size of the plant or concentration (p. 106) in an area.

internal economies the saving in costs made by an industry as a result of increasing the size or scale of plant.

distribution costs the costs of sending goods (↑) to a market area (↓).

---

oil refinery
some products available from crude oil (petroleum)

fuel oil/bitumen
heavy gas oil
gas oil
kerosene/white spirit
naptha/petrol
gas/LPG

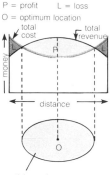

**spatial margins
to profitability**

P = profit   L = loss

O = optimum location

spatial margins
to profitability

market (*n*) the place where goods (↑) are sold.

market area the area which is served by an
industry's goods (↑).

spatial margins to profitability the limits of an
area where revenues are greater than costs
and thus industry is in a profitable location (p. 106).

deglomeration economies the amount which
can be saved by an industry by moving to a
new location (p. 106) away from a large group
of industries, for example in a city.

scale of production the level of unit output
(p. 220) of an industry or plant. The various
sizes are related to economies of scale (↑) in
production and to market area (↑).

value added by manufacture the amount of
money value for manufacturing (p. 118) or
services. It is obtained from the difference
between the value of things paid for outside,
e.g. raw materials (p. 121), and the value of
the finished product (p. 121). It is a measure
of the worth of an industry.

minimum efficient size a measure of the scale
of production necessary for an industry to be
economic (↑) e.g. in the car assembly (p. 121)
industry of the USA the size may be about
800,000 units each year. **MES** (*abbr*).

product mix the range of finished products
(p. 121) made by a firm (p. 112) or industry.

fixed capital one of the factors of production
(p. 124) for industry; it includes land,
buildings and equipment.

capital intensive of an industry which uses
much fixed capital (↑) as part of its factors of
production (p. 124). This may be because of
substitution of capital (p. 114) for labour.

unit cost the average cost of a good (↑) which
takes account of the production costs (↓) in
the manufacture (p. 118) of the product.

venture capital the amount of money provided
for the growth of an enterprise (↑) beyond its
early life. It is very important at that time
because the industry faces an increase in
competition (p. 76).

production costs the amount of money required
to manufacture (p. 118) goods (↑) at a place.

substitution of capital for labour the replacing of some labour by more equipment (*see* fixed capital (p. 123)) to give a more capital intensive (p. 123) industry. The opposite may also take place.

market threshold the smallest market (p. 123) size and lowest demand at which making a good (p. 122) becomes economic (p. 122). This is different for different industries.

integration (*n*) the way in which many parts of an industry may work together in the manufacture (p. 118) of a good (p. 122), e.g. the iron and steel industry. **integrated** (*adj*).

factors of production the main things needed by a firm (p. 112) to produce goods (p. 122). These are mainly labour enterprise (p. 122), capital (p. 114) and land. The price paid for these varies geographically and they are an important part of production costs (p. 123).

opportunity cost a measure of the loss caused by not using a particular resource.

corporation (*n*) an industrial organization which allows various and large scales of production (p. 123). It usually has separate branches for different activities. **corporate** (*adj*).

durable goods the class of manufactured (p. 118) products for which demand can change greatly over a short time, e.g. cars.

non-durable goods the class of manufactured (p. 118) products for which demand changes little over a short time, e.g. food and clothes.

comparative advantage the idea that areas produce goods (p. 122) for which they are better suited than other areas. This leads to regional specialization (p. 179).

co-operative (*n*) an economic system where people have a share in the whole activity, e.g. planned state farming (p. 130) co-operatives in Eastern Europe, Mondragon (p. 183) industrial co-operative. In farming, it may also be a number of farmers who have come together as a large group so that production, buying and marketing (p. 123) may be carried out more effectively. Each farmer, however, still owns his land. **co-operate** (*v*).

agricultural region
Africa

intensive farming and plantations

subsistence farming with low yields

nomadic herding

ranching

land not used for farming

shifting cultivation

crop combination region
Iowa. USA

corn → oats → beans

corn → oats

corn → oats → hay (dried grass) → wheat

wheat → hay → corn → oats → industrial crop → barley

corn → oats → hay → beans → wheat

corn → oats → hay

agriculture (*n*) the use of the Earth's land surface for growing plants or animals that are useful to man. **agricultural** (*adj*).

agricultural geography the study of that part of the Earth's surface whose nature results from agricultural activity. Special attention is paid to the description and explanation of the differences between areas. It is a branch of economic geography.

agricultural region an area of land with its own agricultural character that separates it from other areas. Its character may be expressed through a particular arrangement of crops and animals, or through a simple thing such as area under a certain crop.

farm unit the farm, seen as the main unit of agriculture for the purpose of calculation. For example, there are about 150,000 farm units in England and Wales.

crop combination region an area whose agricultural character is expressed by a commonly repeated order in which various crops are planted over the years. Such an order is not characteristic of the areas near by.

agro-industry (*n*) a union of agricultural and industrial activity. usually in a single farm unit (↑).

agro-town (*n*) a village where up to 10,000 farm workers may live: they travel daily to work in the fields around. Agro-towns are found especially in south Italy, Sicily and Sardinia.

tillage (*n*) the use of land for yearly crops. This is especially important in eastern England.

corn belt that part of east-central United States where the growing of corn (p. 137) is especially important.

alp (*n*) a grassy shoulder or flat area above a valley side in a mountain region (p. 241) (especially the Alps), and which is grazed (i.e. the grass is eaten) by cattle. **alpine** (*adj*).

farm type the nature of a farm unit (↑) as described by the main type of agricultural activity that is followed.

farming system the range of activities carried out on a farm unit (↑).

intensive farming an activity where much capital (p. 114) has been spent on buildings, machines and fertilizer (p. 136) in a relatively small area. Production is therefore high.

extensive farming (1) an activity that covers a large area and so is helped by economies of scale (p. 122). A single crop is often grown, and much capital (p. 114) may have been spent on·machinery. (2) an agricultural activity that is spread out over a very large area because of the poor quality of the land. The yield of a given area is therefore low.

monoculture (*n*) the cultivation (p. 135) of a single crop, often over a large area. This may be to supply the market (p. 123), for example the growing of certain plantation (p. 128) crops in the tropics (p. 241), or as part of subsistence farming (p. 128), for example the sweet potato in the New Guinea Highlands. It may lead to a loss of soil quality.

polyculture (*n*) *see* mixed farming (↓).

arable farming the cultivation (p. 135) of crops such as cereals (p. 137) and vegetables on land that has been ploughed (p. 136).

cereal farming the cultivation (p. 135) of cereals (p. 137). In Britain these cover about three-quarters of the land under tillage (p. 125) and are the main cash crops (p. 140).

market gardening another name for horticulture (↓) which shows the early relation between garden crops and the nearby town market (p. 123).

horticulture (*n*) (1) The production of vegetables in gardens next to houses, especially in the tropics (p. 241). (2) A type of intensive farming (↑), especially of vegetables, fruits and flowers, and carried out either near towns which provide the market (p. 123) for the crops, or in areas which are especially suitable because of soil and climate (p. 7). Also known as **market gardening** or **truck farming**.

truck farming = horticulture (↑). The word 'truck' (lorry) draws attention to the importance of transport.

intensive farming in England and Wales measured by the importance of intensive crops

km
0   160

■ high
□ quite high
▨ quite low
▤ low

market gardening
von Thünen land use zones, changed to show where market gardening occurs

■ city market

■ flower growing

□ truck and fruit farming

▨ dairy farming

dairy farming in England and Wales: the number of cows on 40 hectares of agricultural land

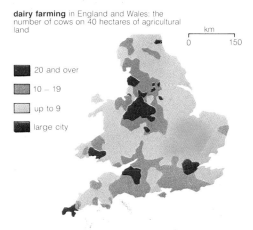

km
0   150

■ 20 and over

▦ 10 – 19

☐ up to 9

▨ large city

mixed farming
a mixed farm in central England

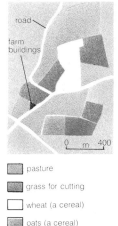

road

farm buildings

0   m   400

▨ pasture

▨ grass for cutting

☐ wheat (a cereal)

▨ oats (a cereal)

▨ barley (a cereal)

dairy farming the keeping of cows in order to produce milk, although meat is also now important. Milk is the main agricultural (p. 125) product in Britain and comes mainly from the western lowlands. The milk is sold as liquid or goes to make cheese and other products.

mixed farming a grouping together of several different types of agricultural activity, usually on the same farm unit (p. 125). For example, mixed farms in Europe often grow grains, root crops and vegetables, together with some animals.

dry farming a way of growing crops in semi-arid (dry) conditions and without irrigation (p. 135). Mulching (p. 135) may be used to prevent evaporation (p. 94). Dry farming usually leads to the loss of valuable top soil by wind erosion (p. 20).

ranching (*n*) a form of extensive farming (↑) in which animals such as sheep, goats and cattle are grown for the market (p. 123) on land which is generally not suitable for arable farming (↑). It is important in the Americas, Australia and New Zealand.

game farming the use of wild life as part of the economy (p. 122) of a farm.

game ranching the use by man of the wild life of
an area. It is carried out on relatively poor
land and over a large area, for example on the
grasslands of East and Central Africa.
Tourism (p. 214) may grow as a result.

pastoralism (*n*) a kind of economy (p. 122)
based on grazing (p. 138) animals which are
used for meat, milk, skins or other products. It
includes both the relatively forward pastoral
economies of Australia and New Zealand, and
also nomadism (↓). **pastoral** (*adj*).

plantation
where plantation agriculture is found

plantation agriculture

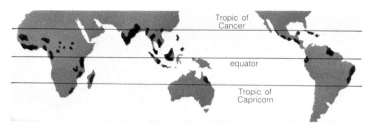

Tropic of
Cancer

equator

Tropic of
Capricorn

plantation (*n*) a large farm unit (p. 125) found
mainly in the tropics (p. 241) which grows crops
for the market (p. 123). It may have machinery
for packing the crops produced. A former
leaning to monoculture (p. 126) is now changing.

peasant farming a kind of farming found widely
in the Third World (p. 184) and characterized
by small farms, lack of capital (p. 114), and low
pay. Only a small amount of food is left over to
be sold. Mixed farming (p. 127), with different
plants in the same field, is common.

subsistence farming a kind of farming where all
that is produced is used. It is now disappearing,
as more and more farms send at least part of
their produce to the market (p. 123).

transhumance (*n*) the seasonal movement of
animals by people who live in a fixed settlement
to pasture (p. 132) elsewhere. In Europe
animals are often driven to high level pastures
in the summer, while in Africa the movement is
related to the wet and dry seasons.

nomadism (*n*) the movement of animals by a people which has no fixed settlement. The animals are moved in a search for new pasture (p. 132), and in Africa this often means travelling to areas where seasonal rain has produced fresh grass. **nomadic** (*adj*).

hunting and gathering a way of life that depends on the hunting of animals for food and useful materials, and on the gathering of natural crops. This way of life has been usual for most of human history, but it is now followed by only a few tens of thousands of people.

holding (*n*) a farm of any size that may or may not be owned by the people who work on it.

smallholding Tanzania, East Africa

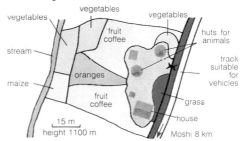

smallholding (*n*) (1) a small farm such as a minifundio (p. 131), often characterized by mixed farming (p. 127). (2) a farm in Britain of about 12 hectares and first set up to provide land for workers who owned none, to settle soldiers after World War I, and to help those out of work in the 1930s.

farm structure the size of a farm and the way its fields are laid out. In many countries of the Third World (p. 184), farms are very small (up to a few hectares) and made up of fields scattered at some distance from each other.

land tenure the arrangement under which land is held. This varies according to the social system (p. 217), ranging from the case where one person owns the land to that where the state (p. 173) owns all. Tenure with a time limit, or limited rights to the land, is quite common.

common land land over which certain people have rights although they do not own it. For example, they may have the right to use it as pasture (p. 132) for grazing (p. 138) animals.

land reform (1) the peaceful sharing-out of land, often brought about by dividing large farms, e.g. latifundios (↓), among agricultural workers who have little or no land. (2) the consolidation (↓) of many small farms.

agrarian reform land reform (↑) together with any social changes that might be necessary, including, for example, improvements in land tenure (p. 129).

agrarian revolution the forced sharing-out of land, often during a time of violent social change, from large land-owners to those who possess little or no land. These have taken place in Haiti (1803) and Cuba (1959–60).

fragmentation (*n*) the result of a continual division of holdings (p. 129). It is common in the older agricultural lands of Europe and Asia, especially in countries where, when a land-owner dies, the land must, by law, be divided equally among the heirs.

consolidation (*n*) the joining of smallholdings (p. 129) (which may have been a result of fragmentation (↑)) in order to make farms of an economic (p. 122) size. It usually takes a long time, for social reasons, but has been quite successful in, e.g. Kenya. **consolidate** (*v*).

remembrement (*n*) a consolidation (↑) of holdings (p. 129) in France, helped by government.

state farm a large holding (p. 129) that is owned and run by the state (p. 173) especially in the USSR. State farms have also been set up in Cuba and Iran, as well as in certain African countries (Ghana, Tanzania), in an attempt to increase food production.

collective (farm) a holding (p. 129) that is owned or lived on by a number of people who work to a plan that has already been laid down. Collective farms are found in the USSR and eastern Europe, and also in Italy, India and Japan. The setting up of collective farms is called collectivization.

collective (farm)
part of a collective farm in Hungary

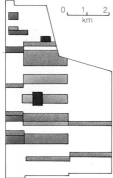

fodder crops

cereals

industrial crops

vegetables

trees

built-up area

latifundio (*n*) a very large holding (p. 129) in a Mediterranean country and elsewhere, especially South America. It is characterized by ranching (p. 127) and by cereal (p. 137) production. **latifundios** (*pl*).

minifundio (*n*) a small farm, often near a latifundio (↑) and belonging to the native people. It is usually too small to support a family. Tension between the two kinds of holding (p. 129) is common in Central and South America. **minifundios** (*pl*).

microfundio (*n*) a very small land holding (p. 129) in Latin America. It is a result of the division of a minifundio (↑) following a growth of population (p. 146) on poor land. The pay of workers may be less than one-third of that provided by a minifundio. **microfundios** (*pl*).

hacienda (*n*) a large farm unit (p. 125) in Spanish Latin America. These were first set up in the sixteenth and seventeenth centuries. In Chile in 1955 a few haciendas controlled 87% of the farm land.

kibbutz (*n*) a co-operative (p. 124) in Israel run by several families who own and work the land between them. It has been especially important in the growth of agriculture in the arid (dry) areas of the country. **kibbutzim** (*pl*).

clearance (*n*) the destruction of vegetation (p. 86) so that crops of value to man can be grown in its place. It is often the first step in the agricultural use of an area.

slash-and-burn

burnt area with stumps set in a clearing

slash-and-burn a way of clearing land for cultivation (p. 135) by cutting down the vegetation (p. 86) and burning it. It is often followed by shifting cultivation (↓).

shifting cultivation a kind of farming characterized by the movement from time to time of both the cultivated (p. 135) area and of the related settlement. It is practised widely in the Third World (p. 184).

bush fallowing = shifting cultivation (↑).

swidden agriculture = shifting cultivation (↑).

fallow (*n*) land that is not producing a useful crop at present. This rest allows it to get back some of its lost quality. **fallow** (*adj*).

rotation (*n*) a system (p. 217) by which different crops, with different demands on the soil, are grown each year. It is a way of avoiding the loss of soil quality that happens when the same crop is grown on a piece of land for several years. It was important in Britain in the nineteenth century, but is not now common because chemical fertilizers (p. 136) are used. **rotate** (*v*).

land rotation a kind of farming practice in which the cultivated (p. 135) area is changed in a regular order. For example, a field may be cropped for one to three years, then left fallow (p. 131) for up to twenty years. Land rotation is common in the tropics (p. 241).

pasture (*n*) (1) a field planted with grass or any other crop that can be eaten by grazing (p. 138) animals. It may always stay as pasture, or may be part of a rotation (↑) system (p. 217). (2) any area where grass grows naturally and provides food for grazing animals. **pastoral** (*adj*).

pasture rotation the movement of grazing (p. 138) animals from one pasture (↑) to another to avoid over-grazing. The electric fence is a way of stopping animals moving between pastures in America, South Africa and Australia.

open range system a kind of pastoralism (p. 128) under which animals can go where they like. As a result they are spread widely in the wet season but they gather around water holes in the dry season.

ley (*n*) land on which grass is grown for only a short time. In eastern England one-year leys are common, while in eastern Scotland leys of two to three years are found. Further west, leys may last for up to ten years.

Norfolk rotation a kind of rotation (↑) under which crops that increase soil quality follow those that lessen it. A field is planted with a grain crop (lowers soil quality) in the first year. This is followed by a root crop (increases soil quality) in the second year, and by another grain crop in the third year. In the next two years grass or legumes (p. 138) are grown for grazing (p. 138) animals whose droppings increase soil quality. The rotation then starts again.

Norfolk rotation

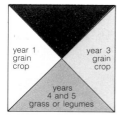

year 1
grain
crop

year 3
grain
crop

years
4 and 5
grass or legumes

open field

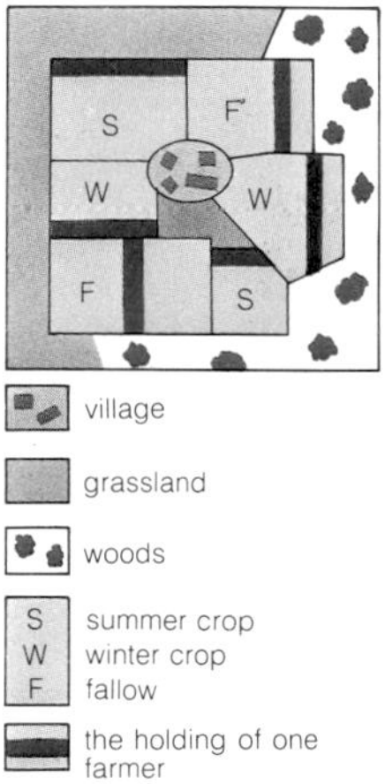
S
F
W
W
F
S
village
grassland
woods
S summer crop
W winter crop
F fallow
the holding of one farmer

two-field system

field 1 (cultivated)	field 2 (fallow)
planted in autumn with winter cereal, e.g. wheat	ploughed at start of summer
planted at start of year with spring crops, e.g. barley	

folding (*n*) the restriction of grazing (p. 138) animals to a small part of a larger area. It was especially used as part of the Norfolk rotation (↑) when sheep were close-folded on fodder crops (p. 138), but it is not now common. **fold** (*v*).

open field a large field near a village in early England and Germany. It was divided into parts, each held by a family. It was cultivated one year in two (*see* two-field system (↓)), or two years in three (*see* three-field system (↓)).

two-field system a kind of rotation (↑) on an open field (↑) in early Britain and Europe before enclosure (p. 134). A field was used for crops one year, and was left fallow (p. 131) the next.

three-field system a kind of rotation (↑) on open fields (↑) and in use in Britain before the agricultural revolution (↓). A field was used for a cereal (p. 137) crop one year, a further cereal the next, and it was left fallow (p. 131) in the third.

agricultural revolution in Britain, the changes in agriculture that took place during the 1700s when the old two-field (↑) and three-field (↑) systems gave way to new rotations (↑) with the cultivation (p. 135) of root crops.

manorial system an arrangement characterized by a large farm called a manor. This was run by its owner (the lord of the manor) as a subsistence farm (p. 128). Village farmers worked the land and supplied the lord with crops. In return they were protected by the lord's soldiers.

three-field system

	field 1 (cultivated)	field 2 (cultivated)	field 3 (fallow)
autumn	plant cereal (wheat)		
spring		plant cereal (barley or oats)	
early summer			plough
late summer	harvest	harvest	plough

enclosure (*n*) the joining of many small pieces of common land (p. 130) and farm land into a larger area that was more economic (p. 122). This began in England around 1200 and continued for several hundred years. It was especially rapid between 1485 and 1603, and from 1761 to 1815. **enclose** (*v*).

coppice-with-standards woodlands in which the *standards* are trees grown for larger sizes of wood, and the *coppice* is underwood which is cut more often for smaller sizes of wood.

royal forest an area, not necessarily wooded, where special laws were in force for conservation (p. 76) of the king's deer.

infield-outfield a system (p. 217) of land use that was common in much of western Europe in medieval times and later, and which is still found in the tropics (p. 241). The infield was the land near the village that was cultivated (↓) all the time; the outfield lay beyond and was left fallow (p. 131) every so often.

ffridd (*n*) an area of relatively poor hill pasture (p. 132) in Wales which is fenced in, and whose quality is between that of open moorland and improved land (p. 137).

combined cultivation a specialized form of agricultural land use in the tropics (p. 241), in which crops are grown for several years between young trees of economic (p. 122) value. The aim is to lessen shifting cultivation (p. 131) which is harmful to trees.

forestry (*n*) (1) land that is covered by trees; (2) the science of the cultivation (↓) of trees.

terrace cultivation the growing of crops on small land areas that have been laid out on steep slopes. The soil is held back by walls up to 15 m high. This system (p. 217) has been used widely, e.g. in the wine areas of western Europe and in the tea plantations (p. 128) of Sri Lanka. A useful side-effect is the prevention of soil erosion (p. 20).

padi field a level piece of land which is used for growing rice. There is usually a bund (↓) around it which holds in the water needed by the rice plants during their growing season.

coppice-with-standards

standard
cut about
every
80 – 100
years

coppice
cut about
every
10 – 15
years

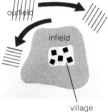

infield-outfield
infield-outfield system

outfield

infield

village

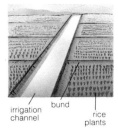

padi field

irrigation
channel

bund

rice
plants

agrotechnical (*adj*) of the use of technology (p. 175) in agriculture, e.g. the use of machine power to do work once done by man or animals.

mechanization (*n*) the use of machines to replace human and animal power in the cultivation ($\downarrow$) of crops, especially in the developed (p. 175) countries. It has been important in cereal farming (p. 126) and especially so in dairy farming (p. 127). **mechanize** (*v*).

cultivation (*n*) the activity of crop growing, including the preparation of the soil. It may need many workers on a small area, as in the tropics (p. 241) or it may be carried out largely by machine, as in much of North America. **cultivate** (*v*).

digging stick a stick with a pointed end, widely used in the tropics (p. 241) for breaking up the soil and for digging holes for plants.

irrigation (*n*) the supply of water, for agricultural purposes, to an area where it is either not present or the amount is too small. It is mainly used in the warm parts of the tropics (p. 241). Here the soil may become salty as a result, and social problems have also arisen. It is also used in the dry areas of Europe and North America. **irrigate** (*v*).

bund (*n*) a bank made of earth which goes round a padi field ($\uparrow$) in Asia. It holds in water supplied by monsoon (p. 71) rains, and is part of an irrigation ($\uparrow$) system. The name is also used for the wall around a lowland reservoir (p. 93).

mulching (*n*) the spreading of vegetable matter around young plants to lessen the rate of evaporation (p. 94). The quality of the soil may also be improved since the amount of humus (p. 90) is increased.

irrigation bund

high land

gully

bund or wall

irrigated area

mulching
as practised by native people in southwest USA

sandy-clay loam

hole open to receive seeds

soil put back and hole filled

loose sand put back to stop capillary movement of water

wind-blown sand

fertilizer the effect of adding fertilizer on crop production

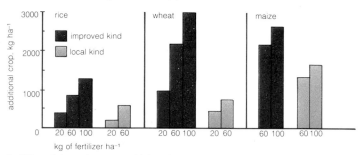

kg of fertilizer ha⁻¹

fertilizer (*n*) a substance which, when added to the soil, makes it yield more crops. It contains three elements that are necessary for plant growth: nitrogen, phosphorus and potassium. Growing world demand has led to the production of man-made fertilizer to add to that naturally obtainable, such as animal waste.

plough (*n*) a sharp-edged instrument for turning over or breaking up the soil. It prepares the ground for cultivation (p. 135), covers the weeds, which make humus (p. 90), and allows frost to break the soil down further. It may be pulled by an animal or a machine. **plough** (*v*).

contour ploughing ploughing (↑) along a level line across a hill slope, so as to lessen the amount of soil erosion (p. 20).

laddering (*n*) a way of lessening the numbers of young rice plants by moving a tool with side bars and cross-pieces (a ladder) across a padi field (p. 134). **ladder** (*v*).

double cropping the planting of a second crop immediately after a first is finished, and in the same place. It is common in the tropics (p. 241) where the wet season lasts for nine months or more, but rare in temperate (p. 241) areas, where it may be part of horticulture (p. 126).

multi-cropping (*n*) the continuous planting and harvesting (↓) of crops with no fallow (p. 131) time between. The different growing-times of the plants, rather than the climate (p. 71), determine the cropping plan.

intercropping (*n*) the cultivation (p. 135) of rows of different crops, often as part of horticulture (p. 126) in the tropics (p. 241), e.g. rice, root crops and vegetables may be grown in different rows. This helps to vary people's food.

agricultural involution the growing of food crops at different levels on a piece of ground (especially in Indonesia). This system (p. 217) allows a large number of crops to be grown in a small area. It is a copy of the way that the plants of the rainforest (p. 87) are arranged.

padi cultivation the cultivation (p. 135) of rice in a padi field (p. 134). The rice may be grown from seed in the padi, or it may be transplanted from a seed bed. It needs many workers, but produces 4 to 6 times as much rice as dry ways of cultivation.

wet rice farming = padi cultivation (↑).

fertile (*adj*) of land or soil which is able to produce a good crop. **fertility** (*n*).

improved land land whose quality has been improved by drainage (↓), by clearance (p. 131), and by the use of fertilizers (↑).

drainage (*n*) the taking away of water from an area that is too wet for agricultural use. Drainage is made possible by the digging of channels (p. 23) in the ground or by laying pipes below the surface.

harvest (*n*) the collection of crops, often by picking or cutting, when they are fully grown and ready for use.

cereal (*n*) a grass plant yielding a grain (also called cereal) that is used for human food. Among the more important cereal crops are rice, barley, maize, oats and wheat.

corn
the corn belt. The area of the USA where corn is important

corn (*n*) (1) in Britain, the grain of cereals (↑) grown for human food. In England it mainly means wheat, in Scotland and Ireland, barley and oats. (2) a plant of the grass family, cultivated (p. 135) especially in the United States, where it is also called maize. It yields a grain used for human food and for making a range of products. The grain can also be fed to animals, as can the plant itself.

maize (*n*) = corn (↑).

rice (*n*) a cereal (p. 137) grass grown for its seed which is the main food of about one-third of the human race. It is grown mainly in the tropics (p. 241). *See* padi field (p. 134) and padi cultivation (p. 137).

legume (*n*) a member (e.g. the bean) of a group of plants able to fix nitrogen from the atmosphere. As a result soil quality is preserved. It is important in rotation (p. 132). *See also* Norfolk rotation (p. 132).

fodder crop a crop which is grown to feed animals. It may be eaten immediately or may be prepared first.

draught animal an animal that is used to carry out a simple task, such as pulling a plough (p. 136) or driving a machine for raising water.

zero grazing a way of feeding animals indoors, in places called stalls, on fodder (↑) and on additional foods. It is widely used in Denmark.

fatstock (*n*) animals that are being fattened on good-quality pasture (p. 132) or on fodder crops (↑) before being sold for meat.

store stock animals that may be grown on relatively poor land, e.g. the hill country of west Britain, before being sent to richer pasture (p. 132) for fattening.

grazing (*n*) the activity of an animal such as a cow, sheep or goat (i.e. a grazing animal) when it eats grass or a fodder crop (↑) for food. **graze** (*v*).

draught animal
at work lifting water

agricultural market
the effect of a
city market on
dairy farming

city

area of
milk production

area of milk and
cheese production

area of
butter production

agricultural market (1) the place where buyers and sellers of agricultural produce meet regularly. (2) the various places to which agricultural produce goes.

Agricultural Marketing Acts laws passed in Britain to help the growth of agriculture. Those of 1931 and 1933 led to the setting up of marketing boards (↓).

marketing boards bodies which were set up in Britain by the Agricultural Marketing Acts (↑) of 1931 and 1933 in order to help the sale of agricultural produce. They aimed especially at helping producers, since buyers were becoming very powerful. Each board has control over production levels and prices. For example, the Milk Marketing Board was set up to make sure that the same price was received by producers for liquid milk and for milk that was used for making products such as cheese. The price was the same everywhere, so helping farmers in distant areas.

producer-retailer a farmer (especially in Britain) who sells a part of his produce in an urban (p. 163) market (p. 123). He has become less important with the growth of easier marketing arrangements.

hill sheep subsidy a subsidy (↓) first given in 1941 to help hill sheep farmers in Britain who had the problem of poor pasture (p. 132) and so could only produce animals of low value.

price support system a way of making sure that the price that farmers receive for their produce is held above that which market (p. 123) forces demand. This may be done by the government for produce to be stored, or giving subsidies (↓) on produce that is sold cheaply to other countries. *See* CAP (p. 140).

subsidy (*n*) money paid to a farmer by a central body, e.g. the government, with the aim of helping production in difficult areas, improving ways of farming, helping long-term change, or increasing the production of certain crops.

monopoly (*n*) the control of a market (p. 123) by a single industry or activity. For example, at the start of the twentieth century the coffee industry of Brazil nearly had a monopoly of the world coffee market.

economic rent the return which can be obtained from farming a piece of land over and above that which can be obtained from a piece of the same size where production is just economic (p. 122).

economic rent a model

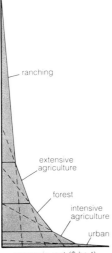

waste

ranching

extensive agriculture

forest

intensive agriculture

urban

economic rent ($ ha⁻¹)

margin of production the distance from a central market (p. 123) at which production of a crop is no longer economic (p. 122) because of the cost of transport or because the crop is easily damaged.

agribusiness (*n*) (1) a large agricultural business which is often part of the activity of a multinational company (p. 113) and which is usually found in a country of the Third World (p. 184). (2) any intensive farm (p. 126) owned by an outside business.

cash crop the crops grown on a farm over and above immediate needs and which are sold. Characteristic of tropical (p. 241) agriculture, where cash crops include coffee and rubber.

share-cropping a system (p. 217) in which the land-owner receives part of the farm income rather than a fixed rent. The farmer often receives 40–50% of the farm income. The system has worked quite well in Australia and New Zealand, and less so in southern USA.

share-farmer a farmer who works the land but does not own it. He usually pays for the use of the land, and may lose his right to work it. This system (p. 217) is common in Europe, Asia, South America and part of southern USA.

Common Agricultural Policy the agricultural aim of the countries belonging to the EEC (p. 180). It makes for agreed prices to EEC producers, as well as a charge on produce from other countries. As a result agricultural production is greater than demand. **CAP** (*abbr*).

green revolution the change brought about through the use of new varieties of high-yielding crops, especially wheat and rice, in the less developed countries (p. 184). A greater use of fertilizer (p. 136) and of irrigation (p. 135) water is necessary.

von Thünen land use zones areas of differing kinds of agriculture, having the form of rings and with an urban (p. 163) market (p. 123) as the common centre. The nature of the farming in each ring partly depends on its distance from that centre. This model (p. 223) was put forward by J. H. von Thünen in 1826.

von Thünen land use zones

■ central city

□ market gardening

□ wood for fires

■ crop farming (no fallow)

□ crop farming, fallow and pasture

■ three-field system

□ animal farming

spatial interaction an idea used by E. L. Ullman (1954). It attempts to show all movements between places, e.g. trade flows, transport, ideas. The amount of movement is usually less over larger distances because of friction of distance ($\downarrow$) or distance decay (p. 142). The principles are complementarity ($\downarrow$), transferability ($\downarrow$) and intervening opportunity ($\downarrow$). *See also* gravity model (p. 142).

friction of distance the idea that concentration (p. 106) of economic (p. 122) activity falls off with distance from a place, e.g. the movement of people and trade will be greater over a short distance than a long distance.

complementary, intervening opportunity and transport flows

+ region with

− region without

- - ► movements (flows)

complementarity (*n*) a relationship between areas such that region (p. 241) 'A' can make goods (p. 122) that are needed by region 'B' It is one of the principles of spatial interaction. **complement** (*v*).

transferability (*n*) one of the principles of spatial interaction which describes the ability of goods (p. 122) to bear transport costs and the effect of these costs on movement. **transfer** (*v*).

intervening opportunity the idea that the amount of movement of, e.g. people and goods (p. 122) between (say) two places is related to the number of opportunities (chances to do things) at places in between. The idea was first used by S. A. Stouffer (1940) for human migration (p. 152), then by E. L. Ullman (1954) as a principle in spatial interaction. Generally the more intervening opportunities there are, the less will be the movement.

diffusion (*n*) the spread of phenomena through space and time. It is used in T. Hägerstrand's (1968) simulation (p. 223) of the spread as waves of new agricultural ideas. **diffuse** (*v*).

space (*n*) the area over which physical and human things are spread. This forms the main part of the study of geography, the other important part is time. **spatial** (*adj*).

accessibility (*n*) the level of ease or difficulty in movement of people or things. **accessible** (*adj*).

phenomenon (*n*) an observable fact or process that can be described scientifically.

centripetal force the name used for those factors which lead to agglomeration (p. 108), for example in the 'pull' of economic (p. 122) activity towards large urban (p. 163) centres.

centrifugal force[2] the name used for those push factors which lead to the dispersal of economic (p. 122) activity from areas of agglomeration (p. 108) to places of less concentration (p. 106), e.g. the movement of industry from crowded inner city (p. 167) areas to the suburban (p. 163) edge

gravity model a model (p. 273) relating a force between two masses to the movement between them, e.g. the amount of migration (p. 152) between two cities is seen as being related to the sizes of the centres and inversely (p. 54) to the distance between them, with less movement the further they are apart.

distance decay the fall in the amount of movement or spatial interaction with increase in distance. This is seen in the gravity model (↑) where flows between places are inverse (p. 54) to the distance. If friction of distance (p. 141) is high, e.g. in mountainous areas, the distance decay will be more rapid.

economic shadow the effect over an area caused by economic (p. 122) activity which lessens the opportunity for such activity in that area and for spatial interaction beyond.

principle of least effort the idea (G. K. Zipf 1949) that people will use the least effort in their behaviour, so that patterns of movement will generally be the least possible. This is similar to the friction of distance (p. 141).

market potential the size of market (p. 123) likely to follow from the location of an industrial plant.

daily urban system the idea of a region (p. 241) with one or more large urban (p. 163) centres whose daily spatial interaction of travel to work and other economic (p. 122) activity spreads over a wide area, often up to 100 miles.

Reilly's law the 'law of retail gravitation' used by W. J. Reilly (1931) to measure the flow of trade from a given place to two towns. It uses the same ideas as the gravity model (↑)

gravity model
expected flows between centres

$$F_{1-2} = a \frac{M_1 M_2}{D_{1-2}{}^2}$$

F = flow

$_{1-2}$ = two centres

M = mass (e.g. population)

D_{1-2} = distance between two centres

a = constant

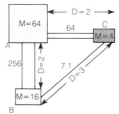

A, B, C = centres

M = mass (population)

D = distance between centres

using the equation, flows between centres = 256, 64, 7.1

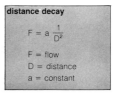

distance decay

$$F = a \frac{1}{D^2}$$

F = flow
D = distance
a = constant

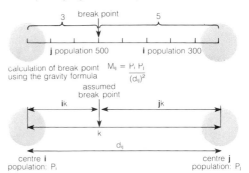

break point between two centres **i** and **j** of population 300 and 500 respectively, eight kilometres apart

calculation of break point using the gravity formula

$$M_{ij} = \frac{P_i \, P_j}{(d_{ij})^2}$$

centre **i**
population: P_i

centre **j**
population: P_j

freight rate
rates as applied to distance by various carriers – road, rail, water

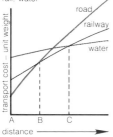

AB   road cheapest
BC   rail cheapest
C+   water cheapest

break point a point between two towns which divides the people who use one centre from those who use another. It is found by use of the gravity model (↑).

freight rate the money paid for loading, moving and unloading materials over given distances. Different rates are given for unit distance for different goods (p. 122); this is important in transferability (p. 141).

free on board a common way to charge for goods (p. 122). The charge is made up of a basic price at point of origin plus the transport cost for movement over a given distance. The cost over distance for a unit of good may be for each mile, or it may be over groups of distance as in stepped rates (p. 144). **fob** (*abbr*).

uniform delivered price a way to charge for goods (p. 122) supplied by adding freight rate (↑) cost to the price as part of the production costs (p. 123), and then to offer the goods at the same price at places at any distance from the source.

basing point price a particular kind of uniform delivered pricing (↑) where the transport of goods (p. 122) is charged as if they had come from a particular place, e.g. Pittsburgh-plus (↓).

Pittsburgh-plus a way of basing point pricing (↑) used until 1926 by the iron and steel industry (p. 118) centred on Pittsburgh, Pennsylvania.

stepped rates

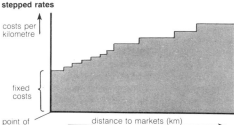

stepped rates a freight rate (p. 143) where cost of moving given goods (p. 122) for each mile is averaged over groups of distances.

phantom freight the cost of transport charged at a place as if the goods (p. 122) had been sent from a given basing point (p. 143).

tapering rates the freight rates (p. 143) which have costs for each mile falling off with increased distance.

blanket rates the freight rates (p. 143) which are charged over a given area for the transport of a particular good (p. 122). Also known as **zonal rates**.

tariff (*n*) the costs added to goods (p. 122) which are imported (p. 113) into a country. The purpose of this may be to lessen the amount of particular goods coming into that country; this is known as a tariff barrier.

terminal charges that part of the costs of the total freight rate (p. 143) which are for handling and loading. Movement costs are line-haul charges.

entrepôt (*n*) a centre, e.g. a port, which gathers in goods (p. 122) for further distribution. From the French word meaning 'storehouse'.

network (*n*) (1) a transport system (p. 217) of fixed infrastructure (p. 113), as in roads etc. (2) line patterns of communication, as in various examples of spatial interaction. Such patterns are seen as being made up of nodes (↓) and edges (↓) in a graph (p. 224). Several indices (p. 224) of networks are calculated by comparing edges and nodes.

tapering rates

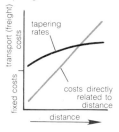

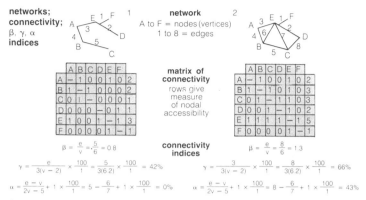

networks; connectivity; β, γ, α indices

network A to F = nodes (vertices) 1 to 8 = edges

matrix of connectivity
rows give measure of nodal accessibility

Matrix 1

	A	B	C	D	E	F	
A	—	1	0	0	1	0	2
B	1	—	1	0	0	0	2
C	0	1	—	0	0	0	1
D	0	0	0	—	0	1	1
E	1	0	0	1	—	1	3
F	0	0	0	0	1	—	1

Matrix 2

	A	B	C	D	E	F	
A	—	1	0	0	1	0	2
B	1	—	1	0	1	0	3
C	0	1	—	1	1	0	3
D	0	0	1	—	1	0	2
E	1	1	1	1	—	1	5
F	0	0	0	0	1	—	1

connectivity indices

$$\beta = \frac{e}{v} = \frac{5}{6} = 0.8$$

$$\gamma = \frac{e}{3(v-2)} \times \frac{100}{1} = \frac{5}{3(6.2)} \times \frac{100}{1} = 42\%$$

$$\alpha = \frac{e-v}{2v-5} + 1 \times \frac{100}{1} = 5 - \frac{6}{7} + 1 \times \frac{100}{1} = 0\%$$

$$\beta = \frac{e}{v} = \frac{8}{6} = 1.3$$

$$\gamma = \frac{3}{3(v-2)} \times \frac{100}{1} = \frac{8}{3(6.2)} \times \frac{100}{1} = 66\%$$

$$\alpha = \frac{e-v}{2v-5} + 1 \times \frac{100}{1} = 8 - \frac{6}{7} + 1 \times \frac{100}{1} = 43\%$$

beta index
connectivity and economic development

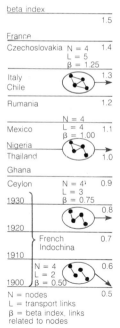

beta index

	1.5
France	
Czechoslovakia	1.4
N = 4 L = 5 β = 1.25	
Italy	1.3
Chile	
Rumania	1.2
N = 4 L = 4 β = 1.00	
Mexico	1.1
Nigeria	
Thailand	1.0
Ghana	
Ceylon N = 4 L = 3 β = 0.75	0.9
1930	
	0.8
1920	
French Indochina	0.7
1910	
N = 4 L = 2 β = 0.50	0.6
1900	0.5

N = nodes
L = transport links
β = beta index, links related to nodes

node (n) a point in a network (↑) which lies at the end of one or more edges (↓). On graphs (p. 224) nodes are called vertices. **nodal** (adj)

vertices see node (↑).

edge (n) a link in a network (↑) which joins up nodes (↑).

connectivity (n) a measure of the movement that is possible in a network (↑) and its directness between points. Various indices (p. 224) can be used to measure the amount of connectivity: for example the beta index (↓), gamma index (↓) and alpha index (↓).

beta (β) index a simple measure of connectivity (↑) which relates the number of edges (↑) to the number of vertices (↑). It is often used as a measure of economic (p. 122) growth where other figures are difficult to obtain.

gamma (γ) index a measure of connectivity (↑) which relates the number of edges (↑) in a network (↑) to the total amount there could be with a given number of vertices (↑).

alpha (α) index a measure of connectivity (↑) similar to the gamma index (↑) but which is related to the number of circuits (↓) in a network (↑) rather than edges (↑).

circuit (n) a path through a network (↑) which begins and ends at the same node (↑) without passing over any edge (↑) more than once.

population density

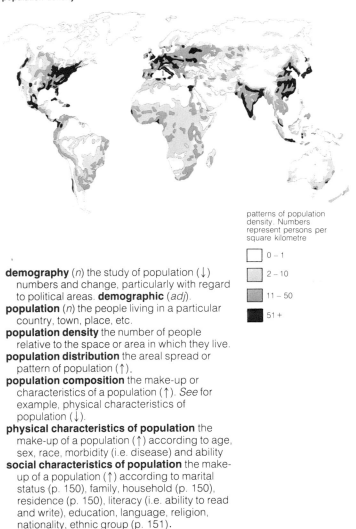

patterns of population
density. Numbers
represent persons per
square kilometre

☐ 0 – 1

☐ 2 – 10

☐ 11 – 50

■ 51 +

demography (*n*) the study of population (↓)
numbers and change, particularly with regard
to political areas. **demographic** (*adj*).

population (*n*) the people living in a particular
country, town, place, etc.

population density the number of people
relative to the space or area in which they live.

population distribution the areal spread or
pattern of population (↑).

population composition the make-up or
characteristics of a population (↑). *See* for
example, physical characteristics of
population (↓).

physical characteristics of population the
make-up of a population (↑) according to age,
sex, race, morbidity (i.e. disease) and ability

social characteristics of population the make-
up of a population (↑) according to marital
status (p. 150), family, household (p. 150),
residence (p. 150), literacy (i.e. ability to read
and write), education, language, religion,
nationality, ethnic group (p. 151).

economic characteristics of population the make-up of a population (↑) according to industry, the kind of work done, and the amount of money obtained for that work.

population dynamics the nature and causes of population (↑) change, particularly as influenced by fertility (p. 151), mortality (p. 152) and migration (p. 152).

population geography the study of areal differences in population (↑) and population characteristics, particularly in relation to physical and human environments (p. 82)

ecumene (*n*) the part of the world's land area which is lived in. It is about 60% of the whole; the rest is said to be non-ecumene.

population density measure the means of showing how a population (↑) is related to the space in which it lives, e.g. crude population density (↓), or the number of people in an area, number of persons to a room, or the size of the agricultural population in a given area.

crude population density the population (↑) of an area, averaged over that area, regardless of the amount of it which is not productive or lived in, i.e. not ecumene (↑) or non-ecumene.

nutritional density the population (↑) of an area in relation to the amount of that area which is farmed.

central tendency a measure of the centrality of average location (p. 106) of population (↑) in the area concerned. *See also* measures of central tendency (p. 226).

population dispersion measure a measure of the degree to which population (↑) is scattered or spread.

population potential sometimes used to express the nearness of people to a particular point, the ease with which they can get to that point or the effect of that point on people at a distance.

population concentration measure the determination of unevenness of population (↑), especially the degree to which a part of a region (p. 241) has a greater share of the region's population than would be expected from area alone. *See* Lorenz curve (p. 148).

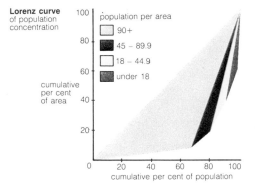

Lorenz curve
of population
concentration

population per area
☐ 90+
■ 45 – 89.9
☐ 18 – 44.9
▨ under 18

cumulative
per cent
of area

cumulative per cent of population

Lorenz curve a population concentration measure
(p. 147). The administrative regions (p. 241)
under study are arranged in descending order
of population density (p. 146), and then
populations (p. 146) and areas of the regions
are added up for each density class. A graph
(p. 224), the Lorenz curve, is then drawn,
plotting the cumulative percentage of area on
the Y-axis and the cumulative percentage of
population on the X-axis.

population spacing measure a measure of the
spacing of population (p. 146) groups such as
villages and towns, e.g. the mean (p. 226)
distance between nearest neighbours.

spatial association measure a measure of
geographical pattern in population (p. 146)
characteristics, e.g. the index of dissimilarity
(↓) and the index of segregation (↓).

index of dissimilarity an index (p. 224) which
describes how two population distributions
(p. 146) relate geographically across a group
of areas. It shows the percentage of the one
population that would need to move into other
areas to give a similar percentage distribution
as the other population.

index of segregation the index (p. 224) of
difference for a sub-group of a population
(p. 146), which shows the degree of
separation or segregation of the sub-group
from the population as a whole.

age indices
dependency ratios in
Mauritius (1959) and the
United Kingdom (1959)

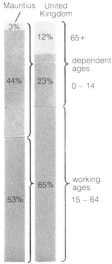

Mauritius   United
            Kingdom

3%   12%   65+

                   } dependent
                     ages
44%   23%   0 – 14

                   } working
53%   65%   ages
                     15 – 64

population pyramid
or age pyramid

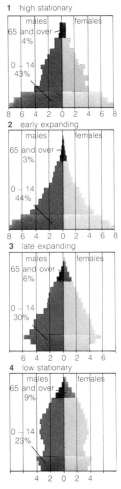

1 high stationary

2 early expanding

3 late expanding

4 low stationary

location quotient[2] the ratio (p. 224) between the percentage of one population (p. 146) in an area and that of a second in the same area. It describes the relative importance of the two populations within a given area. When the quotient, or result of the division, equals 1 (e.g. 80:80), the two are of equal importance.

age structure the make-up of a population (p. 146) according to age, i.e. the age distribution. This may be expressed as age pyramids (↓) or as age indices (↓) for example.

population pyramid a way of showing age-sex data (p. 224) where the Y-axis shows the years from 0 upwards, while the X-axes show either the number or the percentages of the two sexes in each group. Also known as **age pyramid**.

progressive pyramid a population pyramid (↑) with high birth rates (p. 151) and falling death rates (p. 152). The pyramid has a broad base because a large part of the population (p. 146) is young. Such a pyramid is characteristic of many less developed countries (LDCs) (p. 184).

regressive pyramid a population pyramid (↑) with low death rates (p. 152) and decreasing birth rates (p. 151). Such a pyramid is characteristic of developed (p. 175) countries.

stationary pyramid a population pyramid (↑) where there has been little change of birth rates (p. 151) and death rates (p. 152).

age indices indices (p. 224) showing the relationship between the three main age groups, namely children, adults or people who are likely to be self-supporting, and the old or aged. Among the more widely used indices are the old-age index and the dependency ratio (p. 224), i.e. children and aged adults.

differential age structure differences within the age structure (↑) of a country or region (p. 241), especially in relation to different ethnic groups (p. 151) and socio-economic groups (p. 150).

sex composition the make-up of a population (p. 146) according to sex ratio (p. 224). This may be expressed, e.g. as the number of females per 100 or 1000 males (or the other way around), or the percentage in the population.

marital status whether a person is single, married, widowed (the husband or wife has died) or divorced (the marriage has been ended in law). There are three kinds of marriage: monogamy, involving one man and one woman; polygyny, involving one man and two or more women; and polyandry, where a woman marries two or more men.

family[2] (*n*) a social group united by marriage and blood ties, living together (for purposes of characterizing populations (p. 146)).

household (*n*) a group of people living together, and which may or may not be a family.

residence (*n*) a home or living place.

active population this may be described in various ways. Thus the *population of working age* consists mainly of the adult (*see* age indices (p. 149)) population (p. 146). The *working population* is made up of the employed and those looking for work. The *employed population* includes that part of the population actually in work.

work-force = active population (↑).

occupational composition the classification of population (p. 146) according to type of work.

industrial composition the classification of population (p. 146) according to the kinds of economic (p. 122) activity in which the population are employed, usually primary industry (p. 111), e.g. agriculture, secondary industry (p. 111), e.g. manufacturing (p. 118) and tertiary industry (p. 111), e.g. commerce.

socio-economic groups the classification of population (p. 146) according to social and economic (p. 122) considerations, usually based on groupings of kinds of work. In Britain, the socio-economic groups usually recognized are: I professional (e.g. doctors, lawyers); II intermediate (i.e. in between I and III); III skilled; IV partly skilled; V unskilled.

religious composition the classification of population (p. 146) according to religion. Although very important, it is often a difficult classification to make, because of the problems of measuring religious faith and practice

language composition the classification of population (p. 146) according to language. Thus in western Europe alone there are three language families, each with several separate languages: the Germanic (e.g. English); Romance or Latin (e.g. French); and Celtic (e.g. Welsh). Regional (p. 241) variations in the same language, i.e. dialects, may also provide a basis for classification.

ethnic group a population (p. 146) within a larger population which is more or less set apart from it by different racial and social characteristics, especially language, religion and nationality (p. 173).

plural society one with large ethnic groups (↑).

multi-racial society = plural society (↑).

fertility (*n*) the number of live births.

fecundity (*n*) the ability to have children.

generation (*n*) (1) people of roughly the same age group; (2) the time between the birth of a person and the birth of that person's children.

reproduction (*n*) the degree to which people of one generation (↑) give rise to others of the same age in the next generation.

birth rate a measure of a country's fertility (↑), usually expressed as the number of births per 1000 persons per year. The number of births is divided by what is thought to be the population (p. 146) at the middle of the year and the result is then expressed as births per 1000 people. The rate for Britain is 13 per 1000, whereas 40 per 1000 is characteristic of many less developed countries (p. 184).

standardized birth rate the birth rate of a region (p. 241) recalculated on the basis that the age structure (p. 149) of the region is the same as that of the country as a whole.

fertility ratio
of a country

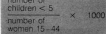
$$\frac{\text{number of children} < 5}{\text{number of women } 15-44} \times 1000$$

fertility ratio the ratio (p. 224) of children to women. This is useful for judging fertility (↑) in countries where there is not enough demographic data (p. 224) for it to be calculated exactly. *See* opposite.

differential fertility the variations in fertility (↑) that arise because of differences in population (p. 146) characteristics, e.g. religion, wealth, etc.

mortality (*n*) the number of deaths.

death rate a measure of a country's mortality (↑) per 1000 persons per year. The number of deaths is divided by what is thought to be the population (p. 146) at the middle of the year, and the result is then expressed as deaths per 1000 people.

standardized mortality rate the death rate (↑) of a region (p. 241) recalculated on the basis that the age structure (p. 149) of the region is the same as that for the country as a whole.

life table mortality rate a table that gives the probability of dying at a particular age, usually calculated on a yearly basis from 0 onwards.

life expectancy (1) the average length of life a person can expect to live; (2) the average length of life that can be expected at a particular age.

differential mortality the variations in mortality (↑) because of differences in population (p. 146) characteristics e.g. sex composition (p. 149), socio-economic groups (p. 150).

migration (*n*) a move from one area, region (p. 241) or country to another, including seasonal movements. Migration may be within a state (p. 173), i.e. internal, or international, i.e. external. **migratory** (*adj*), **migrate** (*v*).

emigration (*n*) migration (↑) from a state (p. 173). **emigrate** (*v*).

immigration (*n*) migration (↑) into a state (p. 173). **immigrate** (*v*).

out-migration (*n*) migration (↑) away from the area in question.

differential migration the tendency for some parts of a population to be more migratory (↑) than others. For example, young men and women are often more likely to migrate than older people in search of employment. Similarly some socio-economic groups (p. 156) are more likely to move than others.

'push and pull' factors 'push' factors are the conditions that cause people to migrate (↑); 'pull' factors are those which influence the choice of the place they move to.

periodic migration there are several kinds of periodic migration. One kind is when a person remains away from the family home for a few years, usually to work for money to send home. Bush-fallowing (p. 131) also includes a form of periodic migration.

rural-urban migration migration (↑) from rural (p. 162) to urban (p. 163) areas. It is the most important kind of internal migration, especially in the early years of industrialization (p. 106). A kind of inter-regional migration (↓).

urban-rural migration the movement of people from cities, conurbations (p. 163) and large towns to settlements in rural (p. 162) areas that are within commuting (daily travelling to work) distance of the larger centres. A kind of inter-regional migration (↓).

intra-urban migration movement from one part of an urban (p. 163) area to another, often to obtain a better quality of life.

inter-regional migration migration (↑) between regions (p. 241), e.g. the westward movement of population (p. 146) in North America in the late 1800s, and the steady movement of people in Britain to London and the south generally.

rate of natural increase a measure of population (p. 146) growth in which migration (↑) is not taken into account, and the amount over time, i.e. the rate, is calculated by taking the death rate (↑) from the birth rate (p. 151), to give the number of people added to each 1000 over the year.

percent natural increase the rate of natural increase (↑) expressed as the rate per 100 rather than the rate per 1000.

annual rate of increase the amount of rate of population (p. 146) growth averaged over a number of years. *See* opposite.

rate of natural decrease a measure of population (p. 146) decrease in which migration (↑) is not taken into account. The amount of rate is calculated by taking the birth rate (p. 151) from the death rate (↑), and the result is expressed as the number by which each 1000 of the population decreases over the year.

annual rate of increase
as calculated by the United Nations

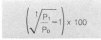

$$\left(\sqrt[t]{\frac{P_1}{P_0}} - 1 \right) \times 100$$

t = number of years
P_0 = population at start of time
P_1 = population at end of time

doubling time the time needed for a population (p. 146) to double in size.

population projections estimates of future population (p. 146) growth and size. They are difficult to make because they depend upon various demographic considerations of fertility (p. 151), mortality (p. 152), marital status (p. 150) and migration (p. 152).

population theories attempts to explain the main characteristics of population (p. 146) growth according to: limiting factors (p. 82), social and economic (p. 122) considerations, especially the demand for people to do work.

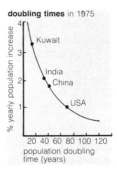

doubling times in 1975

% yearly population increase

Kuwait

India
China

USA

20 40 60 80 100 120
population doubling
time (years)

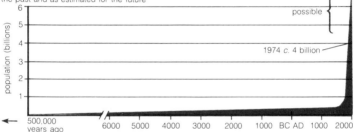

population projections world population growth in the past and as estimated for the future

population (billions)

possible

1974 *c.* 4 billion

500,000
years ago   6000 5000 4000 3000 2000 1000 BC AD 1000 2000

Malthus, T. R. (1766–1834) the English economist who in 1798 produced a theory (↑) which claimed that as the human population (p. 146) can increase geometrically (1,2,4,8...) whereas food production can only be increased arithmetically (1,2,3,4...), population growth will always be held back. This would be done, he believed, by food shortages, war, disease, etc, although he also recognized the importance of late marriage and birth control. Malthus was attacked for the importance he gave to increased death rates (p. 152), and economic (p. 122) development (p. 175) since his time has enabled a much larger world population to be supported. Even so, his ideas are still accepted as sound by neo-Malthusians, who claim that general birth control is necessary.

population policy a course of action planned by a government aimed at slowing or increasing population (p. 146) growth. In less developed countries (p. 184) population policies since 1950 have been directed towards lower growth.

types of population growth the classification of population (p. 146) growth into different types according to the birth (p. 151) and death rate (p. 152) characteristics of the countries.

demographic transition stages

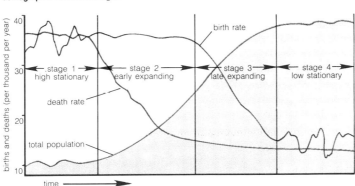

demographic transition for developed countries and less developed countries

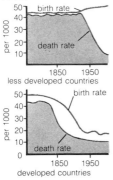

less developed countries

developed countries

demographic transition the drop in fertility (p. 151) in western Europe over the last 130 years or so, which is thought to have been due to economic (p. 122) development (p. 175). However, economic growth on its own has not succeeded in lowering population (p. 146) growth generally in the less developed countries (p. 184) since 1950. In view of this there is doubt whether the idea of demographic transition should be used in relation to LDCs. The transition is often divided into four parts or stages, depending on the birth (p. 151) and death rates (p. 152) of the countries: (1) high stationary; (2) early expanding; (3) late expanding; (4) low stationary.

population cycle = demographic transition (↑)

population/resource ratios measures which attempt to relate a population (p. 146) to its resource base (p. 192), e.g. overpopulation (↓) and underpopulation (↓). Such ratios (p. 224) are highly subjective since resource evaluation (p. 196) depends upon cultural values as well as economic (p. 122) values.

optimum population a population (p. 146) size which, in relation to the area concerned, will allow the best living standards (↓). The view that there is an optimum population has been questioned.

overpopulation a population (p. 146) which is too large in relation to its resource base (p. 192). *Absolute overpopulation* describes the condition where no more economic (p. 122) production is possible. In the case of *relative overpopulation*, greater production is possible but has not yet been obtained.

maximum population an old idea in demography, namely that the land cannot support more people without a fall in living standards (↓). Continuous technological (p. 175) change and economic (p. 122) growth makes it difficult if not impossible to calculate the actual maximum population for any given area. The expression may have more meaning in relation to a subsistence economy (p. 186).

underpopulation a condition in which a population (p. 146) is either too small to use its resource base (p. 192) properly, or in which the resource base could support a larger population without a fall in living standards (↓).

minimum population the smallest number of people necessary for reproduction (p. 151) of the population (p. 146), i.e. about 500. The economic (p. 122) minimum is that size of population below which there are not enough people to do the necessary work.

living standards the levels of comfort in food, housing, and possessions generally, owned by a person or group.

population-resource region a classification of the world into regions (p. 241) according to population/resource ratios (↑).

population/resource ratios
generalized population/
resource regions of North
and South America

 USA type
technology source
low population/
resource ratio

European type
technology source
high population/
resource ratio

Egyptian type
technology deficient
low population/
resource ratio

Brazilian type
technology deficient
low population/
resource ratio

Arctic-desert type
technology deficient
few resources for
human subsistence

central place the hierarchy of central places – W Christaller, based on Southern Germany hierarchy	name of central place	population	distance between centres (km)	size of market area (km²)
1st lowest	Marktort (M)	1,000	7	47
2nd	Amstort (A)	2,000	13	140
3rd	Kreistadt (K)	4,000	21	414
4th	Bezirkstadt (B)	10,000	36	1,243
5th	Gaustadt (G)	30,000	63	3,880
6th	Provinzstadt (P)	100,000	109	11,650
7th highest	Landstadt (L)	500,000	187	35,000

settlement (*n*) a place with buildings for a given population (p. 146). *See* dispersed settlement (p. 162), nucleated settlement (p. 162). **settle** (*v*).

Christaller, Walter (1893–1969) a German economist who determined the most orderly spatial (p. 141) pattern for settlements in his central place theory (↓), put forward in his book *Central Places of Southern Germany* (1933).

central place a point at which entrepreneurs (p. 113) satisfy the demands of a population (p. 146) over a market area (p. 146) or complementary region (p. 159). Each order of place has a particular set of functions which cover various ranges of goods (p. 158) and are placed equal distances apart.

central place theory this attempts to explain a settlement pattern of market centres (p. 158) in a regular order. The distance between centres is related to the range of a good (p. 158) and threshold (p. 221) value. The idea by Christaller (↑) (1933) was for a hierarchy (p. 158) based on (1) marketing principle (p. 159): k = 3, (2) transport principle (p. 159): k = 4, (3) administrative principle (p. 160): k = 7. *See* k number (p. 160). Lösch (1954) changed this by adding manufacturing (p. 118) to the hierarchy to give the Löschian landscape (p. 160).

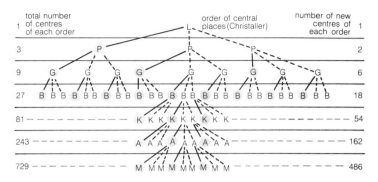

total number of centres of each order	order of central places (Christaller)	number of new centres of each order
1	L	1
3	P P P	2
9	G G G G G G G G G	6
27	B B B B B B B B B B B B B B B B B B B B B B B B B B B	18
81	K K K K K K K K	54
243	A A A A A A A A A	162
729	M M M M M M M M	486

central place hierarchy
number of centres and relation of orders in the k = 3 pattern

hierarchy (*n*) a system (p. 217) of places within given levels of size or importance, and with no places of sizes between these. This would give a clear break between, e.g. hamlets (p. 163) and villages for both size of population (p. 146) and functions. Such a state may be found in small uniform areas. Christaller's (p. 157) and the Löschian landscape (p. 160) are hierarchical sets of central places (p. 157). In large areas, however, a complete range of sizes usually results as described in the rank size rule (p. 161). **hierarchical** (*adj*).

isotropic surface a model (p. 223) land surface where relief (p. 23), population (p. 146), wealth and ease of transport are the same. This 'billiard table' landscape (p. 23) is assumed in models such as Christaller's (p. 157) central place theory (p. 157), Von Thunen's land use model (p. 140) and Weber's industrial location theory (p. 107). Such landscapes are not usually found in reality, but the polderlands (p. 166) of the Netherlands or parts of East Anglia in England may be considered reasonably isotropic.

market centre a central place (p. 157) which serves a given complementary region (↓) with a particular range of a good (↓).

range of a good the greatest distance that people will travel to obtain a particular good (p. 122) or a service.

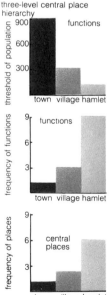

hierarchy
three-level central place hierarchy

functions
(threshold of population: 900, 600, 300 — town, village, hamlet)

functions
(frequency of functions: 9, 6, 3 — town, village, hamlet)

central places
(frequency of places: 9, 6, 3 — town, village, hamlet)

market centre

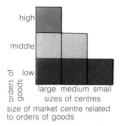

high
middle
low
orders of goods
large medium small
sizes of centres
size of market centre related to orders of goods

order centre settlement status defined by indicator functions

	general store	public house	post office	primary school	butcher	hardware store	clothes store	gents hairdresser	secondary school	shoe shop	bank	supermarket	hospital	cathedral	university
hamlet															
village															
town															
city															

complementary region

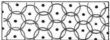

1 complementary regions overlap

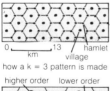

2 how hexagons make the complementary regions

marketing principle
pattern of hamlets and villages service area governed by a k = 3 hierarchy

0 ⌊____13____⌋ hamlet
km village
how a k = 3 pattern is made

higher order   lower order

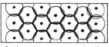

$k = 6 \times 1/3 + 1 = 3$
nesting of centres in k = 3 pattern

higher order   lower order

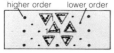

low order centre a central place (p. 157) which serves a small complementary region (↓) with a low range of a good (↑), e.g. bread.

high order centre a central place (p. 157) which serves a large complementary region (↓) with a wide range of a good (↑), e.g. jewellery.

complementary region an area which is served by a central place (p. 157) for particular goods (p. 122) or services. Each order of a good has its own size of region (p. 241). It is also known as the market area (p. 123) and may be similar to the sphere of influence (p. 162) of a town.

'nesting' in Christaller's (p. 157) central place theory (p. 157) this allows the complete control in the hierarchy (↑) by a high order centre (↑) over the next lower order centres (↑) which 'nest' in its complementary region (↑).

marketing principle a k = 3 (see k number (p. 160)) hierarchy (↑) where the trade of a central place (p. 157) is from its own demand and added to by an area with twice its own demand. This comes from one-third of the trade from each of six lower order centres (↑).

transport principle a k = 4 (see k number (p. 160)) hierarchy (↑) where the trade of a centre consists of its own and half that of each of six lower order centres (↑). This results in high order centres (↑) in straight lines which Christaller (p. 160) said would be determined by a need for ease of transport. Also known as **traffic principle**.

administrative principle a k = 7 (see k number (↓)) hierarchy (p. 158) where the trade of a centre consists of its own and the whole of six other lower order centres (p. 159). Christaller (p. 157) said this would be found where central control was the main reason which determined the growth of settlements.

k number a number which, in a hierarchy (p. 158), relates the number of settlements at each level, e.g. in a fixed k = 3 hierarchy the number of settlements is 1, 3, 9, 27, etc. In Christaller's (p. 157) central place theory (p. 157) this figure is fixed through the hierarchy (see fixed k hierarchy (↓)), whilst the Löschian landscape (↓) allows different k numbers to be present in one system (see variable k hierarchy (↓)).

fixed k hierarchy the relation of centres in Christaller's (p. 157) central place theory (p. 157) where the number and distance apart of central places (p. 157) is fixed by a rule such as k = 3, k = 4, k = 7, with the k (↑) value constant through the hierarchy (p. 158).

variable k hierarchy the relation of centres in the Lösch central place theory (p. 157) with a separate range of goods (p. 158), threshold (p. 221) and regional (p. 241) network (p. 144) for each centre. The result is the Löschian landscape (↓) with manufacturing (p. 118) as well as market (p. 123) functions to give 'city-rich' sectors (↓) and 'city-poor' sectors (↓).

urban hierarchy in central place theory (p. 157), the manner in which settlements of a given size are arranged in groups in order of their importance.

Löschian landscape a system (p. 217) put forward by the German economist A. Lösch in 1954 which uses hexagonal (six-sided) market areas (p. 123) but which, unlike Christaller's (p. 157) central place theory (p. 157), allows a variable k hierarchy (↑). Different hierarchies (p. 158) can occur in any one area to give a pattern of 'city-rich' (↓) and 'city-poor' (↓) areas. The pattern is probably truer to reality but it is difficult to use.

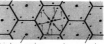

administrative principle
k = 7 pattern with nesting of 6 separate centres in the complementary region of a higher order centre

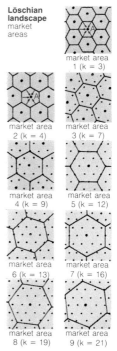

higher order centre   lower order centre

k = 6 × 1 + 1 = 7

Löschian landscape
market areas

market area 1 (k = 3)

market area 2 (k = 4)

market area 3 (k = 7)

market area 4 (k = 9)

market area 5 (k = 12)

market area 6 (k = 13)

market area 7 (k = 16)

market area 8 (k = 19)

market area 9 (k = 21)

Löschian landscape

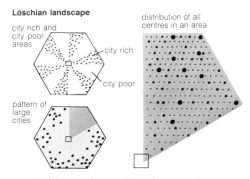

city rich and
city poor
areas

city rich

city poor

distribution of all
centres in an area

pattern of
large
cities

rank size rule

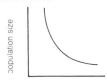

population size

position (rank)

$$P_r = \frac{P_i}{R}$$

P_r = population of rth place

P_i = population of largest place

R = rank of place

rank size pattern for cities in the USA

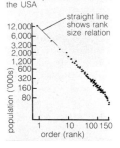

straight line shows rank size relation

population ('000s)

12,000
6,000
3,200
2,000
1,200
600
320
160
80

1   10   100 150
order (rank)

sector (*n*) part of something, for example, a particular economic grouping or area.

'city-rich' sector in the Löschian landscape (↑) a sector (↑) which is obtained by the movement of k networks (p. 144) to give six areas with groups of production centres. The other six areas are 'city-poor' (↓).

'city-poor' sector in the Löschian landscape (↑) a sector (↑) which is obtained by movement of k networks (p. 144) to give six areas with few economic (p. 122) activities or cities. The remaining six areas are city-rich (↑).

hinterland (*n*) a region (p. 241) around a port or other urban (p. 163) centre which is the trade area for that place. A more general description is sphere of influence (p. 162). Other names are nodal or functional region. In central place theory (p. 157) the area is known as a complementary region (p. 159).

rank size rule a rule which says that when cities are placed in order of population (p. 146) size on a logarithmic scale, the graph (p. 224) which results is a fairly straight line. This was first noted by G. K. Zipf in 1949. In practice, if the population of the largest place is known, then the expected population of the Nth rank city will be 1/Nth of the largest. However, this is often not the case in many newer, developing (p. 175) countries the capital (chief city) is larger than expected and is said to be a primate city (p. 152).

primate city a city larger than expected from the rank size rule (p. 161) because of historic factors, e.g. Paris has a population (p. 146) of over 8 million whilst the next largest city, Lyons, has only 1 million. Many less developed countries (p. 184) have such distributions owing to their colonial (p. 174) histories.

sphere of influence the area over which a settlement provides for the needs of goods (p. 122) and services. In central place theory (p. 157), this is called a complementary region (p. 159).

position (*n*) the general place where something is to be found, e.g. the position of New York is on the east coast of North America.

site (*n*) the actual place where something is to be found. For example, the early site of London, England, was on the north bank of the river Thames where it could be crossed by a bridge. **site** (*v*).

rural (*adj*) of areas outside the main urban (↓) centres, with low population density (p. 146) in dispersed settlement (↓) or small towns and villages. *See* rural-urban fringe (p. 167).

settlement pattern the distribution of population (p. 146) groups of different sizes. They may be arranged in a scattered (dispersed) or nucleated (↓) pattern. Attempts to describe settlement patterns describe the distribution but not the function of places.

dispersed settlement a scattered pattern with buildings spread out and without a real centre, as in a hamlet (↓) in a rural (↑) area. On the nearest neighbour index (p. 224) this pattern has a value of greater than 1.

nucleated settlement a pattern with buildings closely grouped together and usually with a central point, as in a village. On the nearest neighbour index (p. 224) this has a value of less than 1.

linear settlement a pattern which follows a line, such as a road or river bank.

city region the area around a large settlement which has close functional ties with the central city.

dispersed settlement

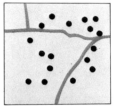

nucleated settlement

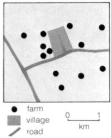

● farm
▇ village
╱ road

0 ——— 1
km

conurbation
West Midlands, UK

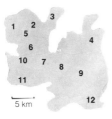

5 km

1 Wolverhampton
2 Willenhall
3 Walsall
4 Sutton Coldfield
5 Bilston
6 Wednesbury
7 West Bromwich
8 Smethwick
9 Birmingham
10 Dudley
11 Stourbridge
12 Solihull

city size distribution the frequency (p. 225) of centres of various sizes. The order may be related to, for example, the rank size rule (p. 161) or to central place theory (p. 157).

suburbanization the movement of people and economic (p. 122) activities such as industry from the central city to outer areas, i.e. the suburbs (↓). **suburbanize** (v).

suburb (n) an area on the edge of the urban (↓) region (p. 241) which is residential (p. 150) but which usually grows into a mix of activities.

urban (adj) of centres of at least 5000 people with various economic (p. 122) and social functions.

conurbation (n) a large and almost continuous urban (↑) area built up from separate nucleated (↑) centres which, owing to urban growth and sprawl (p. 165), have joined.

urbanization (n) the process of increasing population (p. 146) together with the growth of high order (p. 159) functions in urban (↑) areas. It may also mean the increase in population of a country in its largest urban centres.

counterurbanization (n) the scatter of population (p. 146) and economic (p. 122) activity away from the largest urban (↑) centres and beyond the suburbs (↑) to smaller places. See also urbanization (↑).

urban size ratchet the idea that, at a certain size, a city with a developed (p. 175) hinterland (p. 161) is likely to experience continued growth and become more varied in its functions.

corridor (n) the line along which growth of settlement or industry has occurred. This may be urban (↑) sprawl (p. 165) along a transport corridor or planned, as an industrial corridor (p. 116).

classification of settlements various means of grouping places according to (1) their general functions, (2) the structure of their economic (p. 122) activities, (3) by a hierarchy (p. 158) in models (p. 223) such as central place theory (p. 223).

hamlet (n) a very small settlement usually of 10 to 150 people and almost without services or other such functions.

village (*n*) a nucleated settlement (p. 162) usually of 150 to 1000 people with a small number of various functions of low order (p. 159) goods (p. 122) or services.

town (*n*) a settlement usually of 1000 to 2500 people with a number of functions of various kinds, usually including an administrative function from the Town Hall.

city (*n*) a large settlement of 2500 to 500,000 population (p. 146) which possesses a wide range of functions and activities. In Britain it is often defined by having a cathedral.

metropolis (*n*) a very large and complex urban (p. 163) area of 500,000 to 1 million population (p. 146) with high order (p. 159) functions. **metropolitan** (*adj*).

'millionaire city' a name given to those urban (p. 163) areas with over 1 million population (p. 146). It is usually used for the rapidly growing cities of the Third World (p. 184).

megalopolis (*n*) the largest urban (p. 163) areas recognized as having world importance because of their order of functions. The population (p. 146) is usually over 5 million. The name was first used of BOSNYWASH (↓) by J. Gottmann (1961).

BOSNYWASH the regional (p. 241) megalopolis (↑) of the urban (p. 163) corridor (p. 163) in the northeastern United States which stretches from Boston to New York and Washington. It contains about 45 million people.

BOSNYWASH the megalopolis of north-eastern USA

 megalopolis, 1950

recent extensions to megalopolis

metropolitan areas threatened by the exansion of megalopolis

• cores of large cities and towns within or near megalopolis

standard metropolitan statistical area a name
first used in 1960 for metropolitan (↑) areas of
one or more counties in the United States
which have a large population (p. 146), a big
central city and a wide mix of functions.
SMSA (*abbr*).

market town historically, a place with a market
(p. 123) recognized by law to serve a region
(p. 241) around the town. Today such towns have
grown to have many and various functions.

spa (*n*) a place which is noted for its waters which
usually contain particular minerals believed
to give good health. Such places have become
towns for recreation (p. 214). They are named
after the town of Spa in Belgium.

hill town a settlement which has a site of high
relief (p. 23), possibly because it was easy to
defend or to avoid unhealthy lowlands. Often
the agricultural area appears in zones (p. 241)
around the hill. Also known as **hill village**

spring line settlement one of a number of towns
or villages at sites along a line where springs
(p. 98) reach the surface.

ghost town a settlement which usually grew
because of some natural resource (p. 191) such
as minerals but died out when these ended.

sprawl city growth

1 early stages

city 1   city 2
              rural

2 metropolis stage

suburb
              rural

3 megalopolis stage

rural-urban
fringes join

CBD   CBD

☐ homes   ☐ work
CBD = central business district

key settlement a place which because of its
accessibility (p. 141) in the settlement pattern
is chosen as a service centre in rural (p. 162)
planning.

expanded town an urban (p. 163) place which
has been increased in size by planning, in order
to avoid too much growth in larger cities.

sprawl (*n*) the unplanned spread of low density
building at the edges of an urban (p. 163)
area, often in the form of ribbon development
(↓). It is part of a process of decentralization
(p. 114) in urban growth.

ribbon development the sprawl (↑) of urban
(p. 163) land use along a corridor (p. 163) such
as a main transport line out of a central city.

cityport (*n*) an urban (p. 163) area whose growth
has largely depended on port functions and
where trade has led to many industries related
to the port. *See* cityport industrialization (p. 113).

bastide town an historical planned settlement
protected by a castle at a strong point which
could be easily defended, e.g. the bastides of
southwest France. In Britain the name is
usually used for towns such as Caernarfon
(Wales), which grew around a Norman castle.

gap town a settlement sited so as to control the
way through a hill or mountain area. This
makes the place a centre for communications.

frontier town a settlement located (p. 106) at a
certain time on the edge of a populated (p. 146)
region (p. 241). This could be mining town,
agro-town (p. 125), communications centre,
or on a political or other boundary (p. 173).

new town a planned, self-contained settlement
often built to receive the spread of people
from a large urban (p. 163) area.

garden city a completely planned settlement
with much open space, almost self-contained
as a place of decentralization (p. 114).
Ebenezer Howard (1850–1928) used the
idea which led to the world's first garden city
in 1903 at Letchworth, Hertfordshire, England.
The idea was used in planning new towns (↑).

polderland (*n*) low flat land, some below sea-
level, which has been reclaimed from the sea
in the Netherlands. It is protected by banks
and provides new land for farms, e.g.
Wieringermeer Polder of the IJssel-Meer plan
in North Holland.

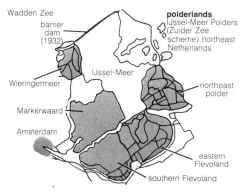

Wadden Zee

barrier
dam
(1932)

Wieringermeer

IJssel-Meer

Markerwaard

Amsterdam

polderlands
IJssel-Meer Polders
(Zuider Zee
scheme), northeast
Netherlands

northeast
polder

eastern
Flevoland

southern Flevoland

green belt
growth of metropolitan green
belt around London

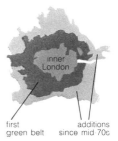

first
green belt

additions
since mid 70s

rural-urban fringe an area at the edge of large
cities with a mix of urban (p. 163) and rural
(p. 162) land uses such as residential (p. 150)
and agricultural. Such areas were called
'rururbia' by Pahl (1966).

gateway city a settlement which controls the
movement of trade from one area to another.
Examples are river or sea ports which serve a
given hinterland (p. 161).

green belt an area of open land where urban
(p. 163) growth is subject to controls. The idea
has been widely used in Britain since the 1947
Town and Country Planning Act. Its purpose is
to keep rural (p. 163) areas around large cities.

rural-urban continuum the idea that the move
from a smaller size of settlement to larger
ones also means a change in the nature of
social structure, with a change from mainly
rural (p. 162) to urban (p. 163) ways of life.

inner city the area close to the centre of a city,
usually of poor, old and decaying houses and
other land uses. This is also the transition zone
(p. 168) of the urban (p. 163) area.

ghetto (*n*) an urban (p. 163) residential (p. 150)
area usually in the inner city (↑) which
contains a particular ethnic group (p. 151).

neighbourhood (*n*) an urban (p. 163) area of a
group of people with particular feeling for the
place and special behaviour in their social
relationships. A similar idea is that of the
urban village.

grid iron pattern a street pattern found in a town
or city, in which the streets run at right angles
to each other. It is common in the United
States, e.g. in Manhattan, New York City,
where the avenues run north-south and the
streets east-west.

urban renewal the attempt to improve the
condition of poor inner city (↑) areas, often
by large-scale urban (p. 163) planning.

industrial zone an urban (p. 163) area which is
zoned (p. 241) for industrial activity. Older
areas are usually around the central business
district (p. 168), but newer industrial areas
are often dispersed in the suburbs.

squatter settlements groups of poor houses
built on land not owned by the people who
move there. They are common in and around
large cities in Third World (p. 184) countries,
where they are known by names such as barrio
(↓) and favella (↓). Also known as **shanty
towns**.

shanty town *see* squatter settlements (↑).

favella (*n*) a name for a squatter settlement (↑)
common around large cities in Latin America.
They are usually found on poor hillside areas
away from the central city where the cost of
living is high.

barrio (*n*) an uncontrolled squatter settlement
(↑), usually of ethnic groups (p. 151) in many
urban (p. 163) areas of Latin America.

bidonville (*n*) a shanty town (↑) such as around
the French ex-colonial (p. 174) town of Tunis.
It means 'oil drum town'.

transition zone an area of the inner city (p. 167)
around the central business district (↓) with a
mix of residential (p. 150) and industrial land
uses. It is an area of crowded living
conditions, with much waste land and people
moving further out. This is one of the zones of
E. W. Burgess's concentric zone model
(p. 170).

central business district the central area of the
inner city (p. 167) where offices and similar
uses are usually tightly grouped. This is
generally the most accessible (p. 141) part
of the city. **CBD** (*abbr*).

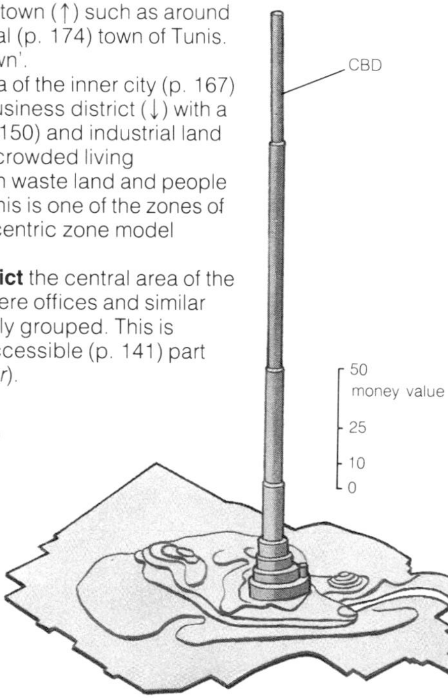

central business district
bid rent increase towards CBD
due to increased competition
for land

CBD

50
money value
25
10
0

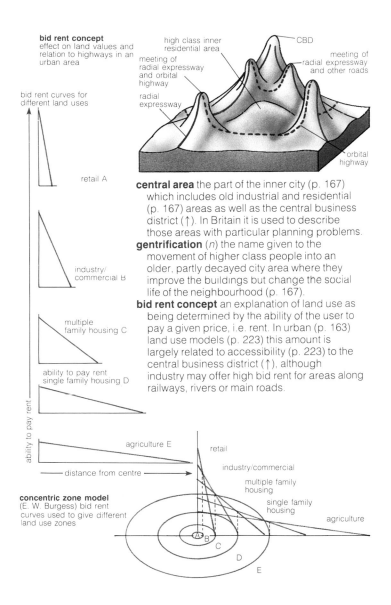

bid rent concept effect on land values and relation to highways in an urban area

bid rent curves for different land uses

retail A

industry/commercial B

multiple family housing C

ability to pay rent single family housing D

agriculture E

distance from centre

high class inner residential area

meeting of radial expressway and orbital highway

radial expressway

CBD

meeting of radial expressway and other roads

orbital highway

concentric zone model (E. W. Burgess) bid rent curves used to give different land use zones

retail

industry/commercial

multiple family housing

single family housing

agriculture

central area the part of the inner city (p. 167) which includes old industrial and residential (p. 167) areas as well as the central business district (↑). In Britain it is used to describe those areas with particular planning problems.

gentrification (*n*) the name given to the movement of higher class people into an older, partly decayed city area where they improve the buildings but change the social life of the neighbourhood (p. 167).

bid rent concept an explanation of land use as being determined by the ability of the user to pay a given price, i.e. rent. In urban (p. 163) land use models (p. 223) this amount is largely related to accessibility (p. 223) to the central business district (↑), although industry may offer high bid rent for areas along railways, rivers or main roads.

models of urban structure

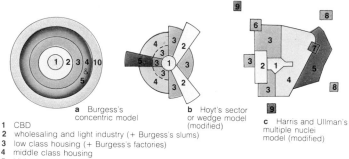

| | a Burgess's concentric model | b Hoyt's sector or wedge model (modified) | c Harris and Ullman's multiple nuclei model (modified) |

1 CBD
2 wholesaling and light industry (+ Burgess's slums)
3 low class housing (+ Burgess's factories)
4 middle class housing
5 high class housing (includes suburbs especially in Burgess's model)
6 heavy industry
7 minor business areas
8 outer suburban housing
9 outer suburban industry
10 wealthy commuter zone (rural-urban fringe)

concentric zone model an attempt by E. W. Burgess (1925) to show urban (p. 163) structure, based on Chicago. Different urban zones (p. 76) grow by competing (p. 241) for limited space by a process of bid rent (p. 76). This model (p. 223) is chiefly based on residential (p. 150) land uses, with little account of industrial uses. The pattern has an inner city (p. 167) central business district (p. 168) giving way to a transition zone (p. 168) and then residential zones.

sector model[1] an attempt by H. Hoyt (1939) to explain patterns of urban (p. 163) land use with areas of different classes of housing along lines from the city centre.

multiple nuclei model a model (p. 223) first used by C. D. Harris and E. L. Ullman (1959). It attempts to explain the scattered pattern of cities which have grown around several points, such as stations, crossroads or existing villages or small towns. Each area tends to come to have specialist functions. Often applied to US cities, e.g. Los Angeles.

ecological urban model an attempt to explain the structure of an urban (p. 163) area by various social factors used in human ecology (p. 72).

historical geography the study of the geography
of earlier times. In particular, the study of the
movements and migrations (p. 152) of peoples.

Palaeolithic Age the 'Old Stone Age', which
ended about 10,000 years ago. Modern man
appeared during this period and developed
basic tools and weapons. **Palaeolithic** (*adj*).

Mesolithic Age the 'Middle Stone Age' of
northwest Europe, which lasted from about
10,000 years ago to roughly 5500 years ago.
Mesolithic (*adj*).

Neolithic Age the 'New Stone Age', which in
northwest Europe began about 5500 years ago,
but which in southwest Asia started about 4000
years earlier. The first farming was done
during the Neolithic Age and this period also
saw the growth of the first settlements.
Neolithic (*adj*).

Bronze Age the first of the metal-working ages,
which in northwest Europe lasted from about
1800 BC to 600 BC.

Iron Age that time when iron became widely
used. It began about 1200 BC in Turkey, but
the knowledge of iron working was not brought
to Britain until about 550 BC.

cultural periods in Britain

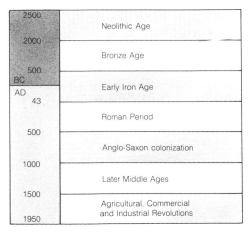

political (*adj*) of public matters, and especially of the way a state (↓) is arranged and governed.

political geography the study that describes and explains the division of the Earth's surface into political regions (p. 241).

electoral geography the study of the degree to which geographical factors may explain the results of elections, especially when these are shown on maps.

sovereignty (*n*) the power of a state over its territory (↓), shown through its ability to make laws which are obeyed by its people. Sovereignty is usually held by the central government. **sovereign** (*adj, n*).

geopolitics (*n*) the study of the way that political power is arranged over the Earth's surface. Its findings have been used in support of the interests of particular states (↓), for example, the territorial claims of Germany in the 1930s. **geopolitical** (*adj*).

heartland concept

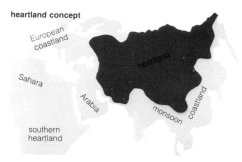

European coastland

Sahara

heartland

Arabia

monsoon coastland

southern heartland

heartland concept the idea, put forward by H. J. Mackinder in 1904, that the central part of Europe and Asia, beyond attack by sea-based states (↓), would act as a core area (↓) for a land-based state that could control the world. In view of present-day weapons this idea has become out-of-date.

core area[1] that area in which a state (↓) has its beginning. For example, the Ile de France (in the Paris Basin) was the core area of present-day France.

core area
France, about AD1100

east border
of French
power,
AD1100

territory (*n*) an area of land that is owned or settled by a group of people. **territorial** (*adj*).

territoriality (*n*) the idea that the basic nature of the state (↓) is found in the land itself, with its boundary (↓) acting as a protection against foreign attack. In view of present-day weapons this idea is now rather out-of-date.

boundary

some types of boundaries in Africa

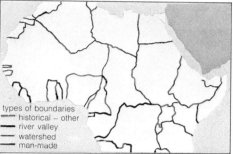

types of boundaries
— historical – other
— river valley
— watershed
— man-made

boundary (*n*) a line separating two things, e.g. on the Earth's surface separating a state (↓) from its neighbours. It also continues above the ground to mark the edge of a state's air space.

frontier (*n*) an area that separates two or more states (↓) and through which a boundary (↑) has not yet been drawn. It is characteristic of a young state pushing into an area where there are few people.

nation (*n*) a large group of people, often united by a common language, land or religion, etc, but more particularly by the desire for common government, peculiar to them. **national** (*adj*).

nationality (*n*) the state of belonging to a nation (↑), or of being a separate nation.

state (*n*) an area of the Earth's surface that is separated from its neighbours by a boundary (↑) or frontier (↑) and whose people live under laws that are laid down chiefly or wholly by the central government.

federal state a state (p. 173) made up of several smaller political regions (p. 241), sometimes also called states, each of which may make its own laws in certain fields, e.g. Canada, Australia and the United States.

unitary state a state (p. 173) in which the central government holds all important political power. There are about five times as many unitary as federal states (↑) in the world. A good example is France.

enclave (*n*) a small area, separate from its homeland but under its political control, and seen from the point of view of the state (p. 173) around it. For example, West Berlin is an enclave when seen from East Germany.

exclave (*n*) a small area inside another state (p. 173) and seen from the point of view of its homeland. For example, West Berlin is an exclave when seen from West Germany.

balkanization (*n*) the division of a former empire or large colony (↓) into several quite small states (p. 173) whose boundaries (p. 173) are often not related to social groupings. Named after the Balkans of southeast Europe.

colony (*n*) a country controlled by a stronger state (p. 173), usually west European, and found in another part of the world. The controlling state is called a 'colonial state'. **colonize** (*v*), **colonial** (*adj*).

condominium (*n*) a country which is controlled by two or more states (p. 173) together, e.g. New Hebrides in the southwest Pacific, where Great Britain shares control with France.

colonialism (*n*) the gaining of control by the coastal states (p. 173) of Western Europe over lands and peoples in other parts of the world. Control by one state over another by economic (p. 122) means is today called neo-colonialism.

decolonization (*n*) the handing over of the colonial (↑) empires of Africa, Asia and the Pacific to their native peoples. This was a rapid process: the empires were at their largest in 1939, while by 1965 decolonization was nearly complete. **decolonize** (*v*).

enclave
Berlin as enclave and exclave

West Germany

West Berlin East Berlin

East Germany

0   100
km

development (*n*) (1) growth; (2) the ability of a national economy (p. 122) to grow at about 5–7% or more each year from a level of little growth; (3) a wider kind of growth, in which the number of poor people is lessened and the problems of lack of work and lack of equality are being met. **developed** (*adj*), **develop** (*v*).

underdevelopment (*n*) a condition in which a country shows less economic (p. 122) growth than might be expected from its resources. About 70% of the world's population (p. 146) lives in a condition of underdevelopment. **underdeveloped** (*adj*).

development economics the study and putting forward of economic (p. 122) and social ideas which, it is hoped, will bring about development (↑) in a country of the Third World (p. 184).

development gap the difference in material production and use between the less developed countries (p. 184) and the developed countries. Some people believe that the development gap is growing wider.

dependency theory the idea that the economic (p. 122) and social systems of the countries of the Third World (p. 184) are not equal to those of the richer countries and so tend to be controlled by them.

sector model[2] the attempt to show regional (p. 241) development (↑) through changes in the importance of areas of the economy (p. 122), with a fall in primary industry (p. 111) and a rise in secondary (p. 111) followed by tertiary (p. 111) and quaternary (p. 111) over time.

technology (*n*) the ability to make or do things, an important determinant of economic (p. 122) development (↑). **technological** (*adj*).

sector model of economic growth

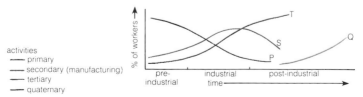

activities
— primary
— secondary (manufacturing)
— tertiary
— quaternary

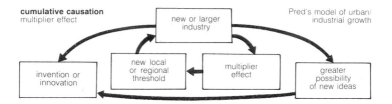

cumulative causation multiplier effect

new or larger industry

Pred's model of urban/industrial growth

invention or innovation

new local or regional threshold

multiplier effect

greater possibility of new ideas

cumulative causation the process used in the model (p. 223) of G. Myrdal (1957) to show how early-favoured areas, probably by initial advantage, grow and develop their economic (p. 122) activity more than peripheral (p. 114) areas. The idea also uses a circular path for the growth as in A. Pred's (1966) model of circular and cumulative causation.

cumulative causation

multiplier growth

multiplier effect the effects that follows from a particular action, such as the opening of a new enterprise (p. 122), which leads to money in the area making demands for other industries and services. It may also lead to linkages (p. 219) as in the regional (p. 241) multiplier.

stages of growth model a model (p. 223) by W. W. Rostow (1971) used to describe the way in which economic (p. 122) and social development (p. 175) of countries can take place over time. There are five steps which lead from low technology (p. 175) to a highly developed industrial economy (p. 122).

stages of growth Rostow's model

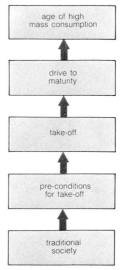

age of high mass consumption

drive to maturity

take-off

pre-conditions for take-off

traditional society

modern industrial society the development (↑) reached by countries in late or post-industrial stages, with a fall in primary (p. 111) and secondary industry (p. 111) and a rise in tertiary (p. 111) and quaternary industry (p. 111).

core[2] (*n*) an area which by early development (p. 175) gains initial advantage and grows more rapidly than the peripheral (↓) region (p. 241), which is dependent on the core spread effects (↓) moving to the periphery.

periphery (*n*) the region (p. 241) in the core-periphery model (↓) which is behind in its development (p. 175) and depends upon the core (↑) area. Backwash effects (↓) return to the core from this region. **peripheral** (*adj*).

core-periphery model
Friedman's development
stages and spatial structure

a pre-industrial

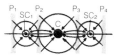

b early industrialization

c industrial maturity

d post-industrial

C = core
P = periphery
SC = secondary core
/ links of spread
and backwash

core-periphery model a model (p. 223) used by
J. Friedmann (1966) to describe the space
economy (p. 122) as related to unequal growth
between regions (p. 241), with the core (↑)
showing most development (p. 175) and with
a periphery (↓) area dependent upon the
core. Spatial interaction takes place from the
core by spread effects (↓) with returns from
the periphery as backwash effects (↓). The
effects can lead to continued polarization
(p. 178) between regions. The model also
describes peripheral regions as either (1)
upward-transition (p. 178), (2) downward-
transition (p. 178), (3) resource-frontier
(p. 178).

spread effects the spatial interactions such as
capital (p. 114) movements from a core (↑) to
peripheral (↑) regions (p. 241). The idea was
used by G. Myrdal (1957) to describe unequal
development (p. 175) between regions. Also
known as **trickle down** (A. O. Hirschman,
1958).

backwash effects the processes of spatial
interaction with the movement of resources from
a peripheral (↑) to a core (↑) area; described
by G. Myrdal (1957). *See* polarization (p. 178).

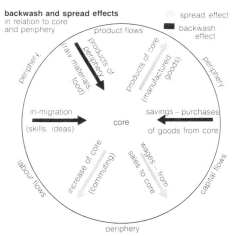

backwash and spread effects
in relation to core
and periphery

polarization (*n*) a name used by A. O. Hirschman (1958) for an increase in uneven development (p. 175) between regions (p. 241). This would come from backwash effects (p. 177) with inequality increased by circular and cumulative causation (p. 176). **polarize** (*v*).

resource-frontier region a peripheral (p. 176) area of new settlement where movement of capital (p. 114) and people into the area makes first use of the natural resources (p. 196), such as agricultural land use or mineral wealth. The idea is used by J. Friedman (1966).

Taaffe stages of growth model

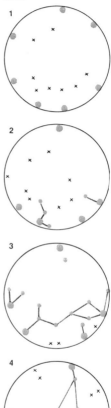

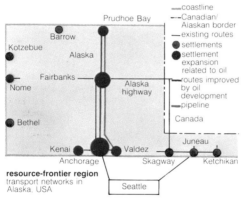

resource-frontier region
transport networks in
Alaska. USA

major settlement
centres — with outside
world contact

x settlements of interior not
linked into urban system

— transport links

downward-transition region a peripheral (p. 176) area of lower economic (p. 122) development (p. 175) because of poor agriculture, loss of minerals or old industry; described by J. Friedmann (1966).

upward-transition region a peripheral (p. 176) area whose closeness to a core (p. 176) area or possession of good natural resources (p. 196) makes for its development (p. 175); described by J. Friedmann (1966). Development corridors (p. 163) may be examples.

Vance mercantile model
stages of spatial
development in a new
country

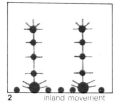

1 scattered ports

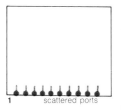

2 inland movement

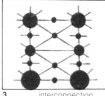

3 interconnection

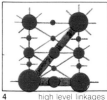

4 high level linkages

  hierarchy of centres

———— transport and other
████ linkages

convergence[2] (*n*) moving together, for example, the path of economic (p. 112) development (p. 175) which leads poorer areas or countries (LDCs (p. 184)) to catch up with richer ones (HDCs). **converge** (*v*).

divergence[2] (*n*) moving apart, for example, the increase in wealth of richer countries, compared to that of poorer countries. **diverge** (*v*).

hollow frontier the movement of people and economic (p. 112) activities over a frontier (border) area, leaving behind a lower density as the frontier continues to move forward.

Taaffe stages of growth model a model (p. 223) used by E. Taaffe (1962) and others to attempt to describe the spatial (p. 141) pattern of development (p. 175) in a generalized island. The four stages depend on trade with the outside world and with building of transport linkages (p. 219) on the island. The stages of development follow a similar path to Rostow's (1971) stages of growth model (p. 223). *See* diagram opposite.

Vance mercantile model a model (p. 223) by J. Vance (1970) which explains settlement patterns (p. 162) in a different way from central place theory (p. 157). It uses historical knowledge to show the economic (p. 122) and spatial (p. 141) growth of settlements as a result of their trade linkages (p. 219), which lead to the spatial interaction of coastal ports and inland centres.

growth pole a name used by F. Perroux (1955) and J. R. Boudeville (1966) to describe an area or centre where a closely related group of industries provides rapid economic (p. 122) growth with a multiplier effect (p. 176) into the general economy.

propulsive industry the leading activity in a growth pole (p. 239).

well-being (*adj*) of the level to which the needs and wants of people are satisfied. The spatial (p. 141) pattern of, e.g. incomes, health and education is the geography of well-being.

regional specialization *see* comparative advantage (p. 124).

common market a grouping of several countries to make a single market (p. 123) for their area. This can then prevent problems over the movement of goods (p. 122), people and capital (p. 114). It also means that all the countries in the group have a common trade policy regarding, e.g. tariff (p. 144) barriers for other countries. An example is the European Economic Community (EEC) (↓).

European Economic Community the group of twelve countries in western Europe which form a common market (p. 123). **EEC** (*abbr*).

Tennessee Valley Authority an organization of the United States government. From 1933 it has worked to control floods (p. 99) and soil erosion (p. 20), to develop (p. 175) electric power, forest and agricultural resources (p. 191), and to support industrial growth in parts of Appalachia around the Tennessee river valley and its tributaries (p. 27). It is important as the first example of integrated (p. 124) river basin development. **TVA** (*abbr*).

depressed region an area of a developed (p. 175) country which is economically (p. 122) below the general national level. This can be related to lack of capital (p. 114) and low use of the means of production.

laggard region = depressed region (↑).

General Agreement on Tariffs and Trade an attempt by western countries to prevent problems over free trade. It was formed in 1948 because in the post-war years many countries were each protecting their own trade with tariffs (p. 144). **GATT** (*abbr*).

structure plan the set of planning ideas which are put forward by law in Britain to suggest the way in which the development (p. 175) of an area should continue. It offers ideas, for example, on land use, transport, and the environment (p. 82) and can use various controls over development.

Economic Development Agency a body of the United States government which is used to help depressed regions (↑) with various development (p. 175) projects. **EDA** (*abbr*).

economic development region multistate regional planning in the USA

 first EDRs 1966–67

 later EDRs

Tennessee valley authority (TVA)

Upper Great Lakes
New England
Pacific Northwest
Old West
Mid-America
Mid-Atlantic
Four Corners
Ozarks
Appalachia
Coastal Plains
South West Border
Mid-South

economic development region a defined large area which is used for regional planning (p. 182). **EDR** (*abbr*).

central planning the strong control of a country's development (p. 175) by central government, especially in socialist states (p. 173). During the time of a particular plan, investment is guided into various parts of the economy (p. 122).

declining region an area which has shown a fall in its economy (p. 122). This may be seen in the closure of firms (p. 112) and out-migration (p. 152) of people as a result of processes opposite to the multiplier effect (p. 176).

development highway system a network (p. 144) of roads made or improved in an area to give better accessibility (p. 141) as a help to economic (p. 122) development (p. 175).

interstate highway a wide, long-distance road which cuts across state boundaries (p. 173) in the USA and joins main metropolitan (p. 164) centres over the country to make up a network (p. 144) of many thousands of miles. In Britain such roads are called motorways, in Germany autobahnen and in France autoroutes. Often built as part of development highway system (↑).

development highway system with interstate highways in Appalachia, USA

 under construction and completed

remainder of planned system

interstate highways

N

population of economic planning regions in the UK. 1978

	region	population ('000s)
A	Northern	3,126
B	North West	6,577
C	Yorkshire and Humberside	4,894
D	West Midlands	5,178
E	East Midlands	3,728
F	South West	4,233
G	South East	16,936
H	East Anglia	1,780
I	Scotland	5,206
J	Wales	2,765
K	Northern Ireland	1,537

regional planning
economic planning regions in the UK

• offices of regional planning councils and boards

regional planning the process of taking actions to arrive at some aim for a particular area. This may be economic (p. 122) or social and may be for a large region (p. 241) or be sub-regional. *See* multistate regional planning (↓).

multistate regional planning the large-scale organization of development (p. 175) activities which include more than one country or several states in a federal nation (p. 174), e.g. the economic development regions (p. 181) such as Appalachia in the USA.

regional policy the policy used by government to give help to those regions (p. 241) which are depressed (p. 180), and often also to lessen development (p. 175) in large growth areas. In Britain this means special help for the development areas (↓).

development areas the parts of a country which have economic (p. 122) problems and may be depressed regions (p. 180), and which are given development (p. 175) money, e.g. to try to bring more industry to the area. In Britain various levels of help are given to (1) development areas, (2) intermediate areas.

modernization (*n*) the social and economic (p. 122) changes which take place as a country or region (p. 241) becomes more advanced in its organization. **modernize** (*v*).

National Environmental Policy Act a law in the USA (1969) to make all organizations of the United States government produce a report on the effects of any development (p. 175) action on the environment (p. 82). **NEPA** (*abbr*). *See also* Environment Protection Agency (↓).

development areas
regional development areas in UK

■ special development areas

■ development areas

■ intermediate areas

Environment Protection Agency the organization which carries out many duties, such as pollution control, of the National Environmental Policy Act (↑). **EPA** (*abbr*).

métropole d'équilibre. (*French*) 'balancing city', a large city area which is used as part of regional (p. 241) policy in France to draw economic (p. 122) activity away from the capital city, Paris. e.g. Marseilles.

métropole d'équilibre
regional centres and their
functional regions in France

Randstad
Netherlands

■ built-up areas
······ limits of conurbations

○ métropoles d'équilibre
• regional centres

Randstad the name for the high-growth region (p. 241) of west central Netherlands. It is the core (p. 176) area of the country.

mezzogiorno (*Italian*) the name, meaning 'mid-day' (sun overhead), for a large problem area of southern Italy. This is a peripheral (p. 176) region (p. 241) of slow economic (p. 122) growth and a depressed region (p. 180) with much regional planning (↑).

Mondragon an area of industrial co-operatives (p. 124) in northern Spain.

growth centre a place considered in regional planning (↑) to have the ability to grow and develop (p. 175) with spread effects (p. 177) to the area around, when infrastructure (p. 113) is developed at the centre.

'seed bed' an area for industry, usually in the inner city (p. 167), with small cheap buildings to allow small firms (p. 112) to grow and develop (p. 175). The idea is also often used in parts of new industrial parks (p. 115).

Third World a group of about 100 countries, found mainly in the low latitudes (p. 238). They are among the least developed (p. 175) in the world but may have large quantities of raw materials (p. 121). Many have been under colonial (p. 174) rule.

North-South (*adj*) of the difference between the developed (p. 175) countries of the northern hemisphere (p. 238) and the less developed countries (↓) are which mainly in the south.

Brandt Report a study (1980) of North-South (↑) relations with especial regard to the problems arising from the difference between rich and poor countries. It is named after Willy Brandt, a former leader of West Germany.

less developed country a country of the Third World (↑). **LDC** (*abbr*).

least developed country one of the 29 poorest countries of the world. **LLDC** (*abbr*).

World Bank a bank whose main aim is to provide money to countries of the Third World (↑) for development (p. 175) purposes. This money comes from the developed countries and from OPEC (p. 187).

New International Economic Order a set of ideas whose aim is to help the countries of the Third World (↑) to improve their trade arrangements with the richer countries. **NIEO** (*abbr*).

First Development Decade the ten years from 1960 to 1969 when the richer nations (p. 173) attempted to increase the help they had been giving to the Third World (↑). The results were less than had been expected.

Second Development Decade the ten years from 1970 to 1979 when the aims of the First Development Decade (↑) were continued. The results were again less than had been hoped for, in part because population (p. 146) growth was faster than economic (p. 122) growth.

United Nations Conference on Trade and Development a conference whose aim is to increase trade between the developed (p. 175) countries and the Third World (↑). Several meetings have been held since 1964 when it was set up. **UNCTAD** (*abbr*).

Development Assistance Committee a part of the OECD (Organization for Economic Co-operation and Development). Its members are from the developed (p. 175) countries of western Europe, the United States, Japan, Canada and Australia, and it aims to supply aid (p. 186) to the Third World (↑). **DAC** (*abbr*).

'big push' strategy an idea of the 1950s that development (p. 175) could best be brought about by a sudden push, rather than in a slower and gentler manner.

commodity agreement an arrangement between the producers and buyers of raw materials (p. 121) and agricultural produce with the aim of smoothing out large changes in price. An example is the International Tin Agreement (1971).

soft loan money which is lent, usually to a country of the Third World (p. 184), at a cost less than the average cost at the time, or without any need for it to be paid back in a certain number of years.

terms of trade the ratio (p. 224) between the amount that a country sells its products for, and the prices it must pay for the material it buys. The terms of trade have got worse for many countries of the Third World (p. 184) in the last few years.

primary producer a country that produces and sells raw materials (p. 121) and agricultural products. For example, Brazil was a primary producer for 100 years after 1822, selling mainly coffee and rubber.

primary sector that part of an economy (p. 122) that is concerned with activities such as the production of food, raw materials (p. 121) and hunting and gathering (p. 129).

colonial economy the system (p. 217) under which the economy (p. 122) of a colony (p. 174) was arranged in favour of the colonial state i.e. a country which rules a colony). For example, the economy of Thailand in the later 1800s depended mainly on rice, rubber, tin and wood, of which only the production of rice was controlled by the native people.

terms of trade

money received from exports	
money paid for imports	
> 1	money flowing into country
< 1	money flowing out of the country

primary sector
examples of primary activities

agriculture

extracting and mining

fishing

forestry

culture system a system (p. 217) by which the Dutch turned Indonesia into a large-scale planned economy (p. 122) during the 1800s. Java especially became an important producer of sugar, coffee and tea for Holland. It led to a pauperization (↓) of the native people.

dual economy a Third World (p. 184) economy (p. 122) in which one or two developed (p. 175) core areas (↓) have large underdeveloped (p. 175) regions around them. This arrangement holds up economic growth because the size of the home market (p. 123) is small. Brazil and Argentina are examples of dual economies.

core area[2] an area within a country where economic (p. 122) growth is taking place much more rapidly than is the case elsewhere in that country. Mexico City (Mexico), Buenos Aires (Argentina) and Santiago/Valparaiso (Chile) are examples of core areas.

subsistence economy an economy (p. 122) in which production consists only of those things necessary to support life. Such an economy is characteristic of many countries of the Third World (p. 184).

enclave economy an economy (p. 122) in which new industry (p. 106) is found side by side with backward native activities. It is characteristic of many countries of the Third World (p. 184).

bazaar economy a system (p. 217) under which economic (p. 122) activity is only on a small scale. It is characteristic of the developing (p. 175) countries of the Third World (p. 184).

newly industrializing country a country of the Third World (p. 184) which is developing (p. 175) rapidly. Many of these are in Southeast Asia, and include the Republic of Korea, Hong Kong, Malaysia, Singapore and Taiwan. In Latin America they include Argentina, Brazil and Mexico. **NIC** (abbr).

aid (n) the help given by richer countries to those which are poor, especially in the Third World (p. 184). Aid may consist of food, medicine and technology (p. 175), as well as money.

Organization of Petroleum Exporting Countries
OPEC oil producers and markets

**Organization of Petroleum Exporting
Countries** a group of Third World (p. 184)
countries which have valuable quantities of oil
and which agree between each other on the
price they should charge. **OPEC** (*abbr*).

moneylender (*n*) a person in Asia, or elsewhere,
who lends money at a high cost, e.g. to a
peasant farmer (p. 128) who can then buy
things such as seed and fertilizer (p. 136).
Moneylenders can cause farmers to lose their
land because their costs become greater
than the money they receive for their produce.

pauperization (*n*) the process by which people
become poorer, especially in the Third World
(p. 184). It has several causes. *Geographical
pauperization* is due to natural processes
such as drought (p. 210) and a decrease in
land fertility (p. 151). *Social pauperization* is
due to the way that a social system (p. 217) is
arranged, e.g. it may favour very small farm
units (p. 125) that are not economic (p. 122).
Economic pauperization takes place when
market (p. 123) forces work against a certain
group. **pauperize** (*v*).

extended family a large family group, including
many relatives, common in Africa and South
Asia. A man may have to support many people
and the high cost of this does not help
economic (p. 122) growth.

joint family = extended family (↑).

apartheid (*n*) a policy of the government of South Africa by which black African people are kept separate from the white people and are made to live in certain areas within the country.

caste system a system (p. 217) of social divisions in India. It is a fixed system, under which people cannot move outside their class. This does not help economic (p. 122) development (p. 175).

Great Leap Forward the grouping of collective farms (p. 130) into people's communes (↓) that took place in China during 1958–9.

people's commune a social group in China that is made up of several collective farms (p. 130) that have been joined. It controls its own agricultural, industrial and commercial matters, and contains on average 20,000 people.

villagization (*n*) the process by which a scattered rural (p. 162) population (p. 146) is grouped into villages as part of a planning arrangement.

intermediate technology the use of a simple machine such as a plough (p. 136) in a country of the Third World (p. 184). It is cheap, its tools are easily made and looked after, and it can give work to local people.

appropriate technology the most suitable technology (p. 175) for dealing with a problem, especially in a country of the Third World (p. 184). The idea of appropriate technology became important because less suitable technologies were often used.

tank (*n*) a way of storing water for irrigation (p. 135) purposes, found especially in India and Sri Lanka. It usually consists of a bund (p. 135) blocking a stream. Elsewhere it may simply be a hollow in the ground.

shadouf (*n*) a machine for raising water from a well (p. 95). It consists of a container at the end of a wooden arm which swings about a fixed point. Various forms are used widely in Africa and Asia.

Persian wheels a system (p. 217) by which water is lifted up by containers fixed to slowly turning wooden wheels. It is used for irrigation (p. 135) in a dry area.

shadouf

Persian wheel

qanat side view of a qanat
system

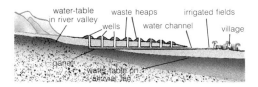

qanat (*n*) a nearly level pipe cut below the ground,
found in the Middle East, especially Iran. It
carries water from a distant aquifer (p. 94) for
irrigation (p. 135) and drinking purposes. It is
about 1.5 m across and may run for 80 km.

Gezira scheme

regional project a system (p. 217) of regional
(p. 241) development (p. 175) carried out by
two or more countries that are near each other
and which stretches across their common
frontier (p. 173), e.g. the Sahel (p. 190)
Programme (covering Senegal, Mauritania,
Mali, Niger, Upper Volta, Chad and Cape
Verde), which aims to improve life in that area.

Gezira scheme an important irrigated (p. 135)
area between the White and Blue Nile rivers
of the Sudan (Gezira is Arabic for 'island').
Water is supplied from a reservoir (p. 93)
behind a dam on the Blue Nile. Cotton is an
important crop in this area.

Volta scheme a large-scale system (p. 217) of
water control on the Volta river, Ghana, West
Africa. Several aims are met: electricity and
irrigation (p. 135) water are provided, and
wet areas are drained (p. 137).

Volta scheme

▬ powerline
╫╫╫╫ railway
▬ road

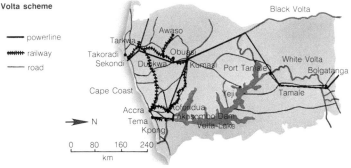

Brasilia project an attempt to develop (p. 175)
the centre-west region (p. 241) of Brazil and
based on the new city of Brasilia.

groundnut scheme an attempt to grow a kind of
nut in what is now southern Tanzania, East
Africa, and then to sell them in western markets
(p. 123). The attempt failed because the cost
was not properly worked out and there were
problems in working the land.

Sahel (*n*) a strip of country along the southern
edge of the Sahara desert, Africa. It is subject
to frequent drought (p. 210). The name means
'shore' in Arabic.

fertile crescent

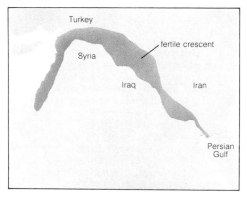

fertile crescent a region (p. 241) of southwest
Asia that includes the valleys of the Tigris and
Euphrates rivers. It is believed to be the area
where agriculture was first practised.

'White Highlands' the high region (p. 241) of
Kenya, East Africa which has a climate (p. 71)
that suits white people. The first land was
settled in 1902, and this later grew to an area
of about 31,000 km² where only Europeans
could own or look after land.

Pan-American Highway a road that runs for
some 45,000 km from northern Mexico to Brazil
and Chile.

behavioural geography the idea that the effect of the environment (p. 82) on people depends partly on whether they see it as a problem or as a resource.

determinism (*n*) the view that the environment (p. 82) controls human activity: it has been said that 'man is a product of the Earth's surface'.

possibilism (*n*) the idea that the environment (p. 82) offers various possibilities, and that the choice between them is made by people.

resource (*n*) a material or non-material thing of value to an organism (p. 73) or to life generally, e.g. mineral resources (p. 196) and intellectual resources.

renewable resources
wind power

non-renewable resources
natural gas

oil

coal

non-renewable resources

greatest production

fast increase as price falls

price rises as production falls

production to date

production

year

renewable resources all living things, provided they are not used or cropped faster than they can produce young and that the habitats (p. 82) are kept in the right condition, e.g. trees.

non-renewable resources resources which cannot be made or re-formed in nature as fast as they are used up, e.g. coal (p. 12).

inexhaustible resources resources like sunlight, and water on the world scale, which will always be present no matter how they are used.

recyclable resources a special kind of non-renewable resource (↑) which can be used over and over again, e.g. many metals.

resource availability the ability of a resource to be used, which will depend on where it is found, the nature and number of the needs for it, and the means that can be employed for obtaining it.

resource base all the resources possessed by a country, or which can be provided for a particular group of organisms (p. 73), purpose or activity.

carrying capacity the number of organisms (p. 73) that can be supported on a given resource base (↑), e.g. the largest number of people that a recreational area can take; it may be controlled by the size of services present, e.g. the number of places for cars.

carrying capacity

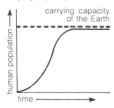

maximum substainable yield MSY
e.g. timber or cattle per unit of area

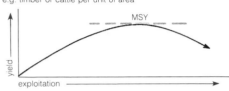

maximum sustainable yield the highest continuous yield that can be obtained from any renewable resources (p. 191). **MSY** (*abbr*).

multiple resource value of a river

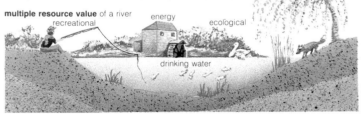

multiple resource value the different values that may be placed on one resource. For example, water resources may provide not only water for drinking, but also ecological resources (p. 196), energy resources and recreational resources. Some values are easier to describe and measure than others.

environmental impact the effect on the environment (p. 82) of (1) increased population (p. 146) and use of resources, especially where these lead to habitat (p. 82) loss and pollution; (2) a hazard or, e.g. large-scale mining.

limits to growth the view, widely accepted, that the human population (p. 146) and industrial production cannot increase continuously because there are not enough resources for them to do so, and because of the greater environmental impact (↑). Instead there will have to be a 'steady state' condition, sooner rather than later, if the world is to avoid serious problems of many kinds. Some people, on the other hand, believe that with further scientific and technological (p. 175) discoveries, much more growth will be possible.

alternative technology technology (p. 175) which differs from that which is common today, in using small quantities of non-renewable resources (p. 191) with the least amount of environmental impact (↑). It also makes fuller use of people, in such a way that they can become more dependent on their own efforts.

environmentalism (*n*) a growing social movement which aims at more and better environmental (p. 82) conservation (p. 76).

resource conservation the conservation (p. 76) of resources, by means, for example, of recycling (p. 195), substitution (p. 195), and the containment of population (p. 146) growth. *See also* limits to growth (↑).

resource development the processes necessary for finding and bringing into productive use a resource or resource base (p. 192).

resource exploitation the obtaining and use of a resource.

resource management the controlled production and use of resources. It should take into account conservation (p. 76) as well as economic (p. 122) considerations.

resource evaluation determination of the nature and value of a resource or resource base (p. 192) for a given area.

land evaluation resource evaluation (p. 193) of land to find the uses – particularly in relation to agriculture, forestry, recreation and ecological (p. 72) conservation (p. 76) – for which it is best suited.

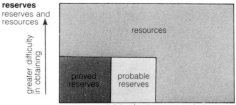

reserves
reserves and resources

greater difficulty in obtaining

resources

proved reserves

probable reserves

greater uncertainty about amount and place ➡

reserves (*n.pl.*) that part of the total stock (↓) of a resource which may be obtained with present technology (p. 175) and under present economic (p. 122) conditions. The reserves may be classed as either *proved* (i.e. measured), or *indicated* and *inferred* (i.e. not proved or measured) where less is known about them. Potential reserves are those which are known but which cannot be obtained for economic and/or technological reasons.

total stock the amount of a resource, in all its various forms, that could be obtained by man. This is usually difficult to measure.

high- and low-value resources groups into which physical resources (p. 196) can be divided, in relation to money value but also place value (↓). Low-value resources have large volume and widespread distribution; the opposite is true of high-value resources.

place value the value that a physical resource (p. 196) has because of where it is found. Thus a low-value resource (p. 191), e.g. sand, has high place value, as it is not economic (p. 122) to move it far, but high-value resources are searched for regardless of distance, and so have low place value, e.g. oil.

static reserve index a measure of the life of a reserve (↑) at present production, i.e. without any allowance being made for changes in the speed of depletion (↓).

exponential reserve index a measure of the life of a reserve (↑) calculated on the basis that growth in production will increase at the same rate as it has done over the last few years, i.e. there is a particular doubling period (↓).

place value

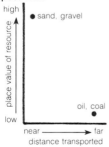

high

place value of resource

low

● sand, gravel

oil, coal
●

near ──────➡ far
distance transported

doubling period the time in which production of a resource increases twofold, as demand increases.

depletion

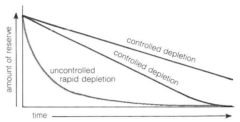

depletion (*n*) the decreasing of a reserve (↑) due to resource exploitation (p. 193), which may be controlled or uncontrolled. Uncontrolled depletion is usually rapid.

substitution

wool

man-made cloth

substitution (*n*) the use of one material resource in the place of another for a particular use or set of uses, e.g. coal (p. 12) for oil in the case of energy substitution.

stockpiling (*n*) the storage of a physical resource (p. 196), to avoid sudden shortages that would arise if supplies were affected by war or serious political problems.

recycling

old car broken up

some parts used to make new steel

made into new car

recycling (*n*) the repeated use of recyclable resources (p. 191). This slows down their depletion (↑) and avoids the pollution that might otherwise arise from their use, although whether recycling is practised depends largely on economic (p. 122) considerations. Copper, glass and paper can all be recycled.

biological resources the general name for genetic resources (p. 191) and ecological resources (p. 196).

genetic resources the various species (p. 81)
of organisms (p. 73) that can be regarded
either now or possibly in the future as
resources. This effectively means all species,
as each may have some use, even if the use
remains to be discovered.

ecological resources these consist of ecosystems
(p. 72) – natural, semi-natural and man-made –
and their various parts. They have multiple
resource value (p. 192) and so their worth can
be expressed in many different ways, e.g.
biological productivity (p. 76), or variety of
habitats (p. 82) and organisms (p. 73).

natural resources the inorganic resources (↓)
and organic resources (↓) provided by the
Earth and the Earth's ecosystems (p. 72).

physical resources the materials which, in
general, can be used to build with, e.g.
physical resources are needed to make roads,
machines and weapons. Land, water, oil, coal
(p. 12) and iron ore are thus all examples of
physical resources. Not included are
resources that come directly from living
organisms (p. 73), such as wood and wool.

material resources = physical resources (↑).

inorganic resources resources (p. 191) formed
by processes and from materials which are not
organic (p. 73), e.g. sand and salt.

organic resources resources provided by living
things – past and present.

primary energy resources renewable resources
(p. 191) of energy, such as radiation (p. 52)
from the Sun, or heat from inside the Earth.

secondary energy resources non-renewable
resources (p. 191) of energy, such as oil and
coal (p. 12).

mineral resources a class of inorganic deposits
(p. 13), consisting of the natural substances of
which rocks are made. They are particularly
important as suppliers of metals.

strategic resources the resources necessary to
support a country during war or time of
economic (p. 122) or political difficulty. They
may be native to the country or result from
stockpiling (p. 195).

non-degradable

plastic or glass bottles

tin cans

pollution (*n*) the presence of energy and/or materials in parts of the ecosystem (p. 72) in amounts and/or forms which cannot be stored and/or used in a harmless form. Pollution results form various kinds of ecological disruption (p. 198). **polluted** (*adj*), **pollute** (*v*).

pollutant (*n*) something which causes pollution, e.g. noise, chemicals. Many pollutants are waste (p. 209) products, which may be in solid, gaseous or liquid (i.e. effluent) forms.

non-degradable (*adj*) of substances, particularly waste (p. 209), that cannot be broken down by decomposers (p. 74) and which therefore increase pollution if not collected and dealt with in the proper way.

non-biodegradable = non-degradable (↑).

biochemical oxygen demand
BOD and human waste in streams and rivers

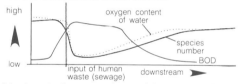

high

oxygen content of water

species number

BOD

low

input of human waste (sewage)   downstream ➤

biochemical oxygen demand the amount of oxygen needed by aerobic (p. 82) decomposers (p. 74) to break down organic matter in water. This is expressed as the number of milligrams of oxygen used per litre of water over 5 days at a temperature of 20°C, after storage in the dark to prevent the formation of more oxygen as a result of photosynthesis (p. 78). Water rich in organic waste (p. 209) will have a high biochemical oxygen demand. **BOD** (*abbr*).

biological amplification
amount of pesticide taken in

carnivore

consumers

herbivore

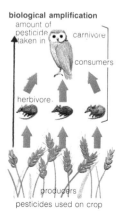

producers

pesticides used on crop

biological amplification the increase in the amount of a substance or substances at each trophic level (p. 74) in a food chain (p. 74) or web (p. 74). Thus a poison or pollutant (p. 197), e.g. a pesticide (p. 201), which may be present in relatively harmless quantities at the first trophic level, may be deadly to a consumer (p. 74) at a higher trophic level. There has been much biological amplification of chlorinated hydrocarbons (p. 200).

eutrophication of a lake

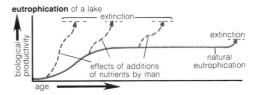

eutrophication (*n*) the addition of nutrients
(p. 76), particularly nitrogen (p. 201) in the
form of nitrates, to an ecosystem (p. 72). This
may happen naturally, but is often speeded
up by man's actions. In water and soils the
results can be particularly harmful. Thus lakes
and rivers, for example, may become oxygen-
poor, since aerobic (p. 82) decomposers
(p. 74) make heavy demands on the supply of
oxygen when dealing with the increased
biological productivity (p. 76) caused by the
additional nutrients.

ecological disruption the accidental and
intended changes due to man in ecosystem
structures (p. 73) and ecosystem dynamics
(p. 76). Such changes may be caused by
pollution and/or may result in pollution.

ecological disruption

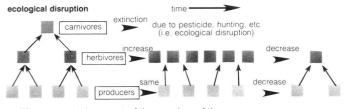

coliform count the count of the number of the
bacterium (p. 77) *Escherichia coli* in water. *E.
coli* is common in the human colon (i.e. the
lower part of the large intestine, which lies
below the stomach). Water purified from
sewage or human effluent (*see* pollutant
(p. 197)) is safe to drink if there are no *E. coli*
present. *E. coli* is harmless but relatively hardy
(strong) and so if it is absent it is probable that
the common disease-causing bacteria are
also absent, for they are less hardy.

without carnivores, the
herbivores at first increase at
the cost of the producers.
Fewer producers will in turn
lead to fewer herbivores

acid rain
the increase of acid rain in
northern Europe from 1956
to 1966

1956

1966

greater
acidity

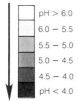

pH > 6.0
6.0 − 5.5
5.5 − 5.0
5.0 − 4.5
4.5 − 4.0
pH < 4.0

pollution control the control, by no means
complete or satisfactory, of pollution by
international and national bodies, e.g.
governments. They decide the different
emission standards (↓) for each kind of
pollutant (p. 197), although many people
believe there is no such thing as a safe limit
for very harmful substances, such as nuclear
waste (p. 209).

pollution monitoring the regular measurement,
against the recognized emission standards (↓),
of the waste (p. 209) put into the environment
(p. 82). The various parts of the
environmental system (p. 217) are also
examined to determine the levels of the
important pollutants (p. 197).

emission standard the limit placed by an
organization, a government or international
body on the amount and/or form and/or way in
which a pollutant (p. 197) can be passed out
into the environment (p. 82). Not all pollutants
are covered by emission standards, and in
any case the standards vary from country to
country. In some parts of the world emission
standards are generally lacking or are not
properly monitored (*see* pollution monitoring
(↑)). For all these reasons pollution control (↑)
throughout the world in general, is still not
satisfactory.

synergistic effect the effect of one pollutant
(p. 197), particularly in the case of pesticides
(p. 201), on another, to produce results that
neither have separately. Thus two pesticides,
each relatively harmless to animals, might
become deadly together.

acid rain the pollution of precipitation (p. 55) by
sulphur and nitrogen-bearing gases, produced
by the burning of oil and coal (p. 12). These
gases and the water in the atmosphere
together yield acids which fall to the ground in
precipitation. Acid rainfall increases leaching
(p. 91) of soils and generally causes
ecological disruption (↑). It can destroy
forests and kill plants and animals in ponds
and streams.

acid rain
the increase of acid rain in
northern Europe from 1956
to 1966

aerosols (*n.pl.*) natural and man-made
substances, e.g. sea salt, dust and smoke,
which, because they consist of very small
particles, float in the atmosphere. Aerosols
become atmospheric pollutants (p. 197) when
their numbers are increased by man's
actions. Possibly they reduce the heating
effect of the Sun and so cause climatic (p. 71)
cooling. They may also increase cloud cover
by providing more centres for condensation
(p. 54) of water, yet rainfall may be lessened
since the average size of droplet is smaller
(*see* collision-coalescence theory (p. 55)).

chlorinated hydrocarbons DDT

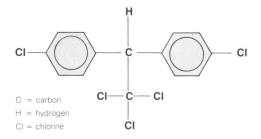

C = carbon
H = hydrogen
Cl = chlorine

chlorinated hydrocarbons a group of man-made
insect poisons or insecticides (*see* pesticide (↓)),
including DDT. They are particularly harmful
to ecosystems (p. 72) because they have an
effect on many different organisms (p. 73), they
remain in the environment (p. 82) for a long
time, they move easily around the environment
and are taken up and stored in animal fats.
organochlorides (*n.pl.*) = chlorinated
hydrocarbons (↑).
detergents (*n.pl.*) cleaning substances rich in
phosphorus, which is a key plant food.
Detergents, together with other wastes (p. 209)
from human society, enter water bodies and
are one of the causes of eutrophication (p. 198).
heavy metals highly poisonous metals such as
mercury, lead and arsenic. They pollute land,
sea and air.

nitrogen a common element (p. 10).
  Substances containing nitrogen, particularly
  nitrates, are important pollutants (p. 197) on
  land, in water and in the air. Some forms of
  nitrogen are key plant foods and so can lead
  to eutrophication (p. 198) if added to the
  environment (p. 82). They are plentiful in
  agricultural wastes and sewage, i.e. human
  waste. Nitrogenous atmospheric pollutants
  arise from the burning of coal and oil. They
  are often harmful to people, but also play a
  part in smog (p. 202) formation. Lastly, waste
  gases from high-flying aeroplanes contain
  nitrogenous substances which can damage
  the ozone layer (p. 52). **nitrogenous** (*adj*).
organophosphates (*n.pl.*) a group of insect
  poisons or insecticides newer than the
  chlorinated hydrocarbons (p. 200). Unlike
  these, organophosphates break down
  quickly and therefore do not remain in the
  environment (p. 82). On the other hand they
  are generally much more dangerous to life.
pesticide (*n*) a substance intended to destroy,
  control or greatly limit a pest or pests, i.e.
  organisms (p. 73) which damage crops or
  lower crop yields. Special kinds of pesticides
  are insecticides for insects, herbicides for
  weeds, and fungicides for fungi (p. 81).
biocide (*n*) = pesticide (↑).
Silent Spring the book by Rachel Carson (1962)
  which first brought to the attention of the public
  at large the risks of pesticide (↑) pollution.
dioxin (*n*) a deadly poisonous substance
  present as an impurity in the pesticide (↑)
  2,4,5,-T. It was dioxin which polluted the soil
  around Seveso, Italy, in 1976.
Minamata disease a kind of poisoning caused
  by organo-mercury fungicides (*see* pesticide
  (↑)), and first noted when it damaged a large
  number of people in Minamata in Japan.
PCBs a class of chlorinated hydrocarbons (↑),
  polychlorinated biphenyls. They are widely
  used in industry and as much as 20% of all
  production may find its way into the environment
  (p. 82) where it can cause serious pollution.

radioactivity (*n*) the natural decay of atoms that gives out much energy in the form of ionizing radiation (p. 52), since it is energetic enough to change atoms it meets into charged particles or ions. If the atoms which are ionized are in an organism (p. 73), the change can result in very harmful effects and even death. Some forms of radiation can pass through walls and travel through air for several hundreds of metres, so that radioactive substances have to be heavily shielded. Radioactivity is measured in curies and the radiation in rads and rems. **radioactive** (*adj*).

thermal (*adj*) of heat. Thermal pollution is waste caused by heat from power stations and industry, e.g. in natural water bodies where it can produce serious ecological disruption (p. 198).

ecocide (*n*) the intentional destruction of the environment (p. 82) by man, e.g. the use of pesticide (p. 201) in the Vietnam war.

ozone depletion the loss of ozone from the ozone layer (p. 52) as a result of the activity of various pollutants (p. 197), e.g. nitrogen (p. 201) and some kinds of aerosols (p. 200) used in aerosol cans. The ozone layer protects life from harmful ionizing radiation (*see* radioactivity (↑)).

smog (*n*) a form of air pollution (↓) which arises when smoke is added to fog. Some atmospheric pollutants (p. 197), namely certain nitrogen (p. 201)-containing substances and smoke from cars, mix under the effect of sunlight to produce ozone (p. 52) gas, which is highly poisonous. This kind of pollution is called photochemical smog.

air pollution pollution of the atmosphere, e.g. by the burning of oil and coal. *See also* acid rain (p. 199), smog (↑) and ozone depletion (↑).

water pollution pollution of natural and man-made water bodies. The pollutants (p. 197) may be chemical or thermal, for example, but eutrophication (p. 198) is the most concern.

soil pollution pollution of the soil, especially with chemical pollutants (p. 197) from agricultural activities, e.g. pesticides (p. 201).

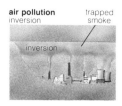

air pollution
inversion   trapped smoke

water pollution
eutrophication

a enough oxygen for aerobes

dead organic matter   input of detergent

b biological productivity of plankton leads to more decay and oxygen usage and also cuts out light falling on plants on bottom

c not enough oxygen for aerobes

floating organic matter

dead organic matter

sensor
aerial mapping camera, with aircraft mount, intervalometer and navigation telescope

remote sensing the science and method of obtaining information about an object or a scene through the analysis (p. 224) of data (p. 224) collected by sensors (↓) that are not in direct contact with the object or scene studied. The information is usually shown in the form of an image (↓).

image (*n*) a pictorial representation of data (p. 224) determined and recorded on film. All 'photographs' taken by aerial cameras (p. 205) on platforms are images, but not all images are photographs since many are recorded by the use of a thermal (↑) or heat-detecting scanner (p. 206). Poor images may be improved by image enhancement (↓). **imagery** (*n*).

image enhancement the use of various methods to improve an image (↑). Enhancement is obtained by increasing the contrast between the objects in the scene and so improving the ease of interpretation of the overall image.

sensor (*n*) an instrument carried on a platform such as an aeroplane or satellite (p. 204) that records radiation (p. 52) from one or more parts of the electromagnetic spectrum (p. 204). Different kinds of sensors are used to record visible, infrared (p. 205) and thermal (↑) images (↑)

spectral sensitivity the ability of a sensor (↑) to pick up wavelengths (p. 205). Individual sensors have fixed limits of spectral sensitivity.

platform (*n*) an aeroplane or satellite (p. 204) carrying the sensor (↑) that gathers remote-sensed data (p. 224).

liquid propellant tank

solid propellant rockets

platform Space Shuttle – a modern platform

payload bay

Spacelab

Orbiter vehicle

satellite (*n*) an unmanned space vehicle used to record information from space or the Earth's surface.

geostationary satellite a platform that is put into orbit at least 36,000 km above the Earth's surface so as to remain in a constant (unvarying) position. The GOES geostationary satellites are in a constant position above the equator (p. 238).

Landsat (*n*) a kind of satellite (↑) used to study the Earth's resources. The Landsat orbits the Earth around the poles (p. 239) with its images (p. 203) recorded mostly by means of a multispectral scanner (p. 207). Landsat images cover regions (p. 241) between 80°N latitude (p. 238) and 80°S latitude. Also known as **ERTS**.

data analysis the process of examining and interpreting data (p. 224) obtained in the form of either pictures or numbers.

electromagnetic energy a kind of energy which may take such forms as visible light, radio waves, heat, ultraviolet light (↓) or X-rays (↓).

Landsat
observation satellite

solar array

data collection antenna

multispectral return scanner beam vidicon (RBV)

electromagnetic radiation divisions of the spectrum

frequency (Hz)	10^{15}	10^{14}	10^{13}	10^{12}	10^{11}	10^{10}	10^{9}	10^{8}

spectrum	UV	visible	near	infrared	microwave	VHF

wavelength → 0.4 0.7 | 1.5 µm   1 mm commonly used   0.8 m
µm µm   radar bands

sensors:
film in cameras
radar/passive microwave detectors
solid state detectors in scanners and radiometers

1 µm (micrometre)
= 1000 mm (nanometres)
= 10^{-6} (metres)

electromagnetic radiation the radiation (p. 52) of electromagnetic energy (↑) from the Earth's surface in various wavelengths (↓) which may be sensed and gathered by sensors (p. 203) carried on platforms. *See also* spectral resolution (↓).

electromagnetic spectrum the complete range of electromagnetic energy (↑).

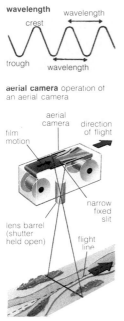

wavelength

crest

trough

aerial camera operation of an aerial camera

aerial camera
direction of flight
film motion

narrow fixed slit

lens barrel (shutter held open)

flight line

X-rays (*n.pl.*) a form of electromagnetic energy (↑) with a very short wavelength (↓).

ultraviolet radiation a form of electromagnetic energy (↑) which occurs on the short wavelength (↓) side of the visible light spectrum.

infrared radiation a form of electromagnetic energy (↑) which occurs on the long wavelength (↓) side of the visible light spectrum.

wavelength (*n*) the distance from one wave peak to the next wave peak.

spectral resolution a name used to describe the position and width of a band of electromagnetic radiation (↑) sensed by instruments on a platform.

aerial camera a machine for taking aerial photographs (↓) from an aeroplane. There are several kinds: one that produces one or more images (p. 203) at the same time, another that takes a continuous image as the aeroplane flies, and one that forms an image of a very wide area.

aerial photograph an image (p. 203) of the Earth's surface or atmosphere produced from a film taken in an aerial camera (↑). There are several kinds, depending on the spectral sensitivity (p. 203) of the film used.

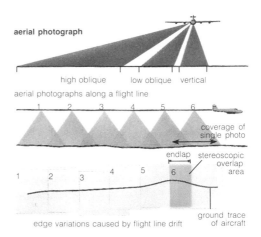

aerial photograph

high oblique   low oblique   vertical

aerial photographs along a flight line

coverage of single photo

endlap   stereoscopic overlap area

ground trace of aircraft

edge variations caused by flight line drift

atmospheric effects effects caused by gases
and particles in the atmosphere which may
absorb electromagnetic radiation (p. 204)
given off from objects on the Earth's surface.
This reduces the energy that reaches the
sensor (p. 203). Gases and particles may
also produce their own radiation which will also
have an effect on the sensor output. Effects
related to the scattering of particles may
therefore result in a 'haze' or mist that reduces
the sharpness of the image (p. 203).

blackbody radiation curves for Earth surface features (at 27°C)
and flowing lava (at 1100°C)

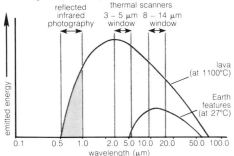

blackbody radiation the radiation (p. 52) of
energy by an object called a blackbody,
which acts as an ideal radiator, i.e. it absorbs
completely, and gives out again, all the
energy that reaches it. The energy it radiates
is a function of its surface temperature. The
distribution of energy from a blackbody varies
with temperature and the blackbody radiation
may be shown on a graph (p. 224) as a curve.
data acquisition a process that depends on
energy sources that can be sensed through
the Earth's atmosphere and the interaction of
energy with Earth surface features. The data
(p. 224) is obtained through sensors (p. 203)
carried in the air or in space. and gathered in
either photographic or numerical form.
scanner (*n*) an instrument for the detection of
images (p. 203).

multispectral scanning the process of collecting information from a number of spectral bands at the same time. Multispectral scanners (↓) use electronic energy detectors which record information from ultraviolet (p. 205) wavelengths (p. 205) and reflected and thermal (p. 202) areas of the spectrum.

multispectral scanner a scanner (↑) that separates incoming energy into several spectral components that may then be sensed as single pieces of information.

photogrammetry (*n*) a technology (p. 175) that is used to obtain measurements and make maps from photographs. This is done through measurements and geometric principles and the use of various equipment, such as glass scales, electronic digitizers (↓) and comparators (a machine which compares images (p. 203) and notes the differences between them). Photogrammetry is mainly used in the preparation of topographic (p. 22) maps.

electronic digitizer an electronic machine with a keyboard, microprocessor and magnifying 'pen' used in the rapid and accurate study of aerial photographs (p. 205).

colour infrared film a film that records green, red and infrared (p. 205) scene energy. It presents the viewer with a 'false-colour' image (p. 203) in which objects that reflect chiefly red energy are recorded as green, whilst infrared reflections are recorded as red images.

aerial thermography a form of remote sensing that is concerned with measuring the variation in temperatures of Earth surface features. It is very useful in the control of Earth resources.

isotherm maps maps produced by thermal (p. 202) scanning (↑) methods. They record the temperature distributions on the Earth's surface.

fiducial marks the marks on the sides of an image (p. 203) that define the area or frame from which spatial (p. 141) measurements are made on aerial photographs (p. 205).

principal points spots on aerial photographs (p. 205) that mark the place where lines drawn from opposite fiducial marks (↑) cross each other.

fiducial marks

fiducial mark

hazard (*n*) a risk of danger. **hazardous** (*adj*).

hazard prediction the attempts made by scientists to predict (i.e. foresee) future catastrophic (↓) events. Prediction methods may include the use of measuring instruments and/or monitoring (watching and noting) the behavioural patterns of animals which may act as a precursor (↓).

risk assessment a way of judging the degree of damage that an area may experience as a result of an earthquake (↓), volcanic eruption (↓) or any other possible hazard.

triggering mechanism an event such as a heavy rainstorm, volcanic eruption (↓) or earthquake (↓) that may cause a flood (p. 99) or landslide (p. 21). A number of triggering mechanisms may act over a length of time to produce a possible hazard.

catastrophe (*n*) an event that causes great suffering and damage or destruction, such as earthquakes (↓), floods (p. 99), landslides (p. 21) or volcanic eruptions (↓). **catastrophic** (*adj*).

earthquake (*n*) a natural movement or vibration (shaking) of the Earth's crust (p. 8) produced by the folding (p. 133) or fracturing (breaking) of Earth material. Earthquakes are common at plate margins (p. 15) and are often linked (p. 219) with volcanic (p. 40) activity.

seismology (*n*) the study of earthquakes (↑).

seismograph (*n*) an instrument for recording the location (p. 106), depth and strength of an earthquake (↑).

epicentre (*n*) a point on the Earth's surface that is centred directly above the focus (↓) of an earthquake (↑).

focus (*n*) *see* hypocentre (↓).

hypocentre (*n*) the point within the Earth at which an earthquake (↑) starts. It is also called the focus and is usually found at depths less than 70 km (43.5 miles). Shallow focus earthquakes start above 70 km, whilst deep focus earthquakes begin below 300 km (186.4 miles).

Richter scale a logarithmic scale used to measure the strength of earthquakes (↑).

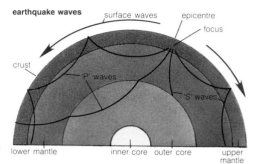

earthquake waves

surface waves · epicentre · focus

crust

'P' waves

'S' waves

lower mantle · inner core · outer core · upper mantle

volcanic eruption
Hawaiian type height of
(shield volcano) eruptive column

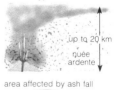

very low

area affected by ash fall

less than 0.1 km²
strongly cone building

Strombolian type

up to 1 km

area affected by ash fall

up to 5 km²
strongly cone building

Plinian type

up to 20 km

nuée ardente

area affected by ash fall

up to 1000 km²
strongly sheet building

earthquake waves shock waves, a form of energy produced by the fracture (breaking) of Earth materials. Waves that pass through the inside of the Earth are called body waves; the energy that reaches the surface makes surface waves.

volcanic eruption the sending out of volcanic (p. 40) material from a volcanic vent (hole) or fissure (crack). Each volcano is noted by its own kind of eruption; some, like those of the Hawaiian volcanoes, are quiet with fluid lavas; others, like that of Mont Pelée (Martinique), are violent with a cloud of gas, stream and hot volcanic debris (dust and pieces of rock) called a nuée ardente (↓).

nuée ardente (*French*) a fiery cloud of volcanic (p. 40) debris that may rapidly cause much damage and destruction. For example, in 1902 when Mount Pelée (Martinique) erupted, the glowing cloud of dust, gas and steam destroyed the town of St Pierre.

waste (*n*) the by-products or rubbish (unwanted materials) produced by man in the use of the Earth's resources and the manufacture of goods (p. 122).

nuclear waste (*n*) the radioactive (p. 52) by-product of nuclear power industries (p. 106) or the mining of radioactive minerals (↓).

radioactive mineral a mineral that gives out energy as radiation (p. 52).

precursor (*n*) an event or change that happens before an earthquake (↑) or volcanic eruption (↑) and which is a sign of the hazard to come.

chemical precursor *see* precursor (p. 209). A change in the temperature of the Earth's crust (p. 8); increased activity of hot springs (p. 98); or an increase in gas from a volcanic vent (hole).

geophysical precursor *see* precursor (↑). A change in the state of the Earth's crust (p. 8), including minor earthquakes.

behavioural precursor *see* precursor (↑). A change in the way animals act, many becoming more nervous or behaving unusually before an earthquake (p. 208) or volcanic eruption (p. 209) .

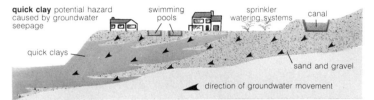

quick clay potential hazard caused by groundwater seepage

swimming pools

sprinkler watering systems

canal

quick clays

sand and gravel

direction of groundwater movement

quick clay a clay (p. 12) or soil that can change quickly from a solid to a liquid state if shaken. Quick clay may move suddenly to cause a catastrophe (p. 208).

flash flood a sudden movement of water and possibly earth materials due to sudden heavy rainfall.

avalanche (*n*) a rapid down-slope movement of snow or ice in mountainous areas. The force of the avalanche varies in relation to the humidity (p. 54) of the snow. Wet snow avalanches are very powerful and may cause widespread damage. Dry snow avalanches create shock waves, and buildings in their pathways are destroyed as a result of changes in air pressure.

drought (*n*) a long time without rain. In temperate (p. 241) areas the official meaning of a drought is when no rain falls for 15 to 20 days. Elsewhere, in arid areas, a drought may last several years. It is especially characteristic of the edges of tropical (p. 241) deserts (p. 87) such as the Sahara, Africa. The Sahel (p. 190) is an area where drought often occurs.

medical geography the study of the relationship between disease (↓) in man and the environmental (p. 82) factors that may have caused it.

welfare geography a branch of geography concerned with the study of how the quality of life varies between different areas. It may include ideas for changing present conditions.

World Health Organization a body of the United Nations set up in 1948 to improve world health conditions. It makes arrangements for the control of various important diseases (↓), and has largely succeeded in the case of malaria. **WHO** (*abbr*) .

malnutrition (*n*) a lack of enough food to support an active life. About 500 million people suffer from malnutrition. This is about one-fifth of the population (p. 146) of the Third World (p. 184)

disease (*n*) a change in bodily behaviour which may have an effect on the survival (length of life) of a creature in its environment (p. 82).

Bilharziasis (*n*) a disease (↑) caused by a kind of soft creeping animal (a worm) that enters the body It is picked up from water that contains snails in which the worms live, and is especially common in the tropics (p. 241). Irrigation (p. 135) may give rise to this disease, as happened in Egypt. Up to 200 million people around the world may suffer from it, especially in Africa. Also known as **schistosomiasis**.

vector (*n*) a carrier of a disease (↑) which is then passed to another animal. The habitat (p. 82) of the vector may determine the location (p. 106) of the disease, e.g. the vector of the disease malaria is usually found only in areas of fresh water in the tropics (p. 241).

anopheles a flying insect. It is a vector (↑) for the disease (↑) malaria, which is the most widespread of all the tropical (p. 241) diseases. Also known as **mosquito**.

plague (*n*) a disease (↑) which killed half the population (p. 146) of Europe in the fourteenth century when it was known as the Black Death. As the Great Plague it again killed many in 1665. The vectors (↑) were rats.

tsetse
main cattle areas and main areas of tsetse fly disease in
Africa

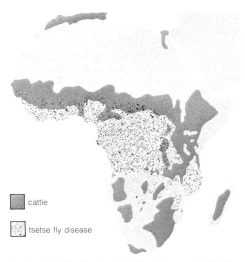

☐ cattle

☐ tsetse fly disease

tsetse (*n*) a fly that carries the disease (p. 211)
   sleeping sickness in tropical (p. 241) Africa.
deficiency disease a disease (p. 211) which
   comes about when man's diet (food) is not
   varied enough. Substances important for
   health are therefore missing. The disease
   kwashiorkor is an example, and is due to a
   shortage of protein in the food eaten in rice-
   growing areas, especially southeast Asia.
trace element a metallic element such as copper
   which, in small amounts, is a necessary part of
   man's diet (food). Too little or too much may
   lead to disease (p. 211) and sometimes
   death. Trace elements may be taken up by
   crops from the soil.
carcinogen (*n*) a substance that produces
   cancers (harmful growths) in animals. Certain
   carcinogens are made by man for use in the
   food industry, e.g. artificial (i.e. non-natural)
   colourings.

occupational disease a disease (p. 211) caused by conditions at work, e.g. the chest disease silicosis has for long been related to the digging out of coal (p. 12). It may start quite quickly, or may not appear for many years.

stress disease a disease (p. 211) which comes about when environmental (p. 82) pressures are too great. For example, it may be caused by overcrowding in an urban (p. 163) area, or by social pressures.

endemic[2] (*adj*) of a disease (p. 211) that is always present in a population (p. 146). For example, the disease cholera appears to be endemic in the Celebes, Indonesia.

pandemic (*adj*) of a disease (p. 211) that spreads over a very wide area, perhaps the whole world.

environmental hazard a pressure placed on man by his environment (p. 82). It may be due to climatic (p. 71) factors, e.g. high or low temperatures, to social factors, e.g. the growth of industry, or to biological factors.

thermal stress the harmful effects of very high and low temperatures on the human body. Low temperatures often lead to diseases (p. 211) of the chest; high temperatures to diseases of the bowel.

windchill (*n*) the cooling effect of air movement and low temperature. It is a better guide to the feeling of cold than temperature alone because it measures about 80% of body heat loss.

medical revolution the improvement in medical care that is taking place in the Third World (p. 184). A result is that populations (p. 146) are increasing in size and in the number of young people.

Clean Air Acts laws passed in Britain in 1956 and 1958 which lessened the amount of pollution (p. 197) that was allowed to be produced, especially from chimneys. They were a result of serious smog (p. 202).

senescent disease a disease (p. 211) due to old age rather than to environmental (p. 82) factors, e.g. some kinds of heart failure.

recreation (*n*) the pleasurable activities carried out by a person during the times when he or she is not working. **recreational** (*adj*).

recreation geography the study of the arrangement in space of recreational activity.

landscape evaluation an attempt to measure landscape (p. 22) quality in objective ways with the aim of comparing the recreational values of different areas. It gives rise to problems, such as the personal nature of the measurement and the difficulty of comparing countries.

tourism (*n*) travel either within a country or in others and with a recreational aim in mind. It produces much wealth for poor countries, but may damage the way of life of the people.

National Trust a state body whose aim is to preserve the countryside and to help people to enjoy it. It owns about 175,000 hectares, divided between farmland and woodland. It was set up for England, Wales and Northern Ireland in 1895, and for Scotland in 1931.

Countryside Acts laws passed in Britain in 1967 and 1968 which tried to make land in the country more open to people. Under the Acts local governments could buy land for use as Country Parks (↓), car parks, and recreation points. Access agreements (↓) were also made under these Acts.

Countryside Commissions bodies set up in Britain under the Countryside Acts (↑). They attempt to lessen the pressures on the countryside through Access Agreements (↓) and through the setting up of recreation points and Country Parks (↓).

National Park an area of fine scenery where human activity is carefully controlled and where the environment (p. 82) is preserved or improved for the enjoyment of people and the conservation (p. 76) of native plants and animals. The first was Yellowstone in the USA, set up in 1892. There are ten National Parks in England and Wales, covering an area of about 12,950 km². This is 10% of the whole area.

access agreement an agreement made between a land-owner in Britain and the governing body in that area which allows the public to use the land for recreation (↑).

Area Of Outstanding Natural Beauty an area of fine countryside in England or Wales, which is used for outdoor recreation. It is usually smaller than a National Park (↑) and includes more improved farmland. There are 28 such areas, covering about 11,655 km². Examples are the Gower (189 km²), the Cotswolds (1,507 km²) and the Shropshire Hills (777 km²).

Country Park a recreation area set up quite close to centres of population (p. 146) in Britain, with the aims of lessening pressures on more distant areas; cutting travel time, and making roads less crowded. Country Parks were made possible by the Countryside Acts (↑).

National Forest Park an area of forest in Britain that is owned by a state body (the Forestry Commission) and used for recreation as well as for producing wood. The forest roads are used for walking, and places are provided for cars, for camping, and for watching animals and birds. It may be quite large: Snowdonia Forest Park (Wales) has an area of 9639 hectares.

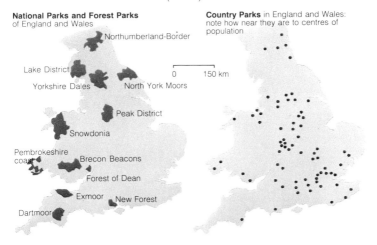

National Parks and Forest Parks of England and Wales

Northumberland-Border

Lake District

Yorkshire Dales

North York Moors

0   150 km

Peak District

Snowdonia

Pembrokeshire coast

Brecon Beacons

Forest of Dean

Exmoor New Forest

Dartmoor

Country Parks in England and Wales: note how near they are to centres of population

Nature Conservancy Council a public body in
Britain whose aims are to look after a collection
of habitats (p. 82) and their characteristic
plants and animals, and to give advice on
matters related to the environment (p. 82).

resort (*n*) a town, usually on the coast, which
owes its rise and importance to tourism (p. 214).
Its growth may have started by accident, but
considerations such as distance from
population (p. 146) centres are important.

Enterprise Neptune an attempt by the National
Trust (p. 214) in the 1960s to buy parts of the
British coast so that they could be preserved
for recreation. By 1974, 255 km had been
bought increasing the Trust's holding to 544 km.

Heritage Coast a length of British coast that has
been recognized by the Countryside
Commission (p. 214) as being of scenic value,
e.g. the North Yorkshire and Cleveland
Heritage Coast, which runs for 55 km between
Saltburn-by-the-Sea and Scalby Ness.

long-distance footpath one of a number of
paths for walkers that were set up in Britain after
1945 and which cover a longer distance than
the old, short paths of the lowlands, e.g. is
the Pennine Way (402 km). In most cases
access agreements (p. 215) were necessary.

'honeypot' area an area that pulls in a large
number of tourists (p. 214), e.g. Tarn Hows,
the most visited beauty spot in the English
Lake District, and Swallow Falls, in the Snowdonia National Park (p. 214) in Wales. The
number of visitors may give rise to problems.

second home a home whose owner lives
elsewhere, and which is used for recreation
for only part of the year. It is much more
common in North America, France and
Scandinavia than in Great Britain.

Wilderness Area a part of North America that
remains unchanged by man, has fine scenery,
and has been recognized by the Wilderness
Act of 1964 and is protected by law.

game park an area where wildlife is preserved
under natural conditions. Game parks are
important to tourism (p. 214) in East Africa.

game parks
of East Africa

Kenya
Mt. Kenya
Aberdares
Nairobi
Olorgesailie
Bubasco
Gedi
Tsavo
Nakuru

Tanzania
Serengeti
Ngordoto
L. Manyara
Mikumi
Ruaha

system (*n*) a grouping of things or ideas which relate to each other to form a whole. In geography, a part of the world that is separate enough to be studied on its own and whose parts are related.

general systems theory the theory (p. 222) first put forward by L. von Bertalanffy in 1937, that a common frame, using systems, could be employed in all the sciences. The theory said that studies of living forms only made sense when they were examined as a whole or as a system (↑). This idea was used in other areas of science, and then the separate systems were united to form general systems theory.

open system a system (p. 217) that allows both matter and energy to pass across its borders. An example is the drainage basin (p. 26) which accepts energy (e.g. the Sun) and matter (e.g. water) and sends out heat and sediment (p. 11). Most natural systems are open.

closed system a system (↑) that allows the input (p. 220) or output (p. 220) of energy but not of matter, e.g. the Earth, as it receives energy from the Sun but little else.

morphological system a system (↑) made up of related elements that can be measured. An example is a beach (p. 46) morphological system made up of the links (p. 219) between elements such as slope, average grain size, range of grain size, and beach firmness.

cascading system a related set of subsystems (p. 219) through which energy or mass passes. Found at various levels, from the Sun's energy flow, down to the small example of a wave moving from deep to less deep water.

morphological system
beach system made up of three related variables

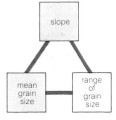

cascading system weathering cascade

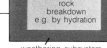

I = input
O = output

water and various substances

rock breakdown e.g. by hydration

weathering subsystem

products of weathering

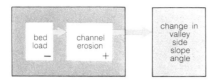

process-response system
showing how a lessening of
bedload increases channel
erosion and changes valley
side slope angle

$+$ = increases
$-$ = reduces

process-response system a system (p. 217)
   formed when morphological (p. 217) and
   cascading systems (p. 217) are joined, e.g. a
   hillslope system in which the morphological
   element (shape) is related to the cascading
   element (movement of water and material).
   The link ($\downarrow$) often demands negative feedback
   (p. 220) which leads to inner changes as
   inputs (p. 220) and outputs (p. 220) change.
control system the system (p. 217) that forms
   when man changes the action of a process-
   response system ($\uparrow$), often by changing an
   important variable ($\downarrow$). An example is the
   system that results when a river meander
   (p. 25) is straightened.
white box system a system that is studied by
   producing the greatest possible knowledge
   about the way its inner structure behaves when
   certain inputs (p. 220) are made.
black box system a system (p. 217) whose
   inputs (p. 220) and outputs (p. 220) are known
   but whose inner order is not understood, e.g.
   the movement of water through a system when
   only rainfall and runoff (p. 96) are known.

control system
the 19th-century
straightening of the Rhine
upstream of Mannheim: an
example of a control system
not in equilibrium

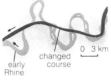

black box system

$\mathbf{I}$ = input $\mathbf{O}$ = output

white box system

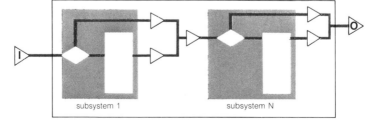

subsystem (*n*) a small system (p. 217) with input (p. 220) and output (p. 220) which is part of a larger system. For example, a morphological system (p. 217) may be part of a process-response system (↑).

variable[1] (*n*) a part of a system (p. 217) which, when acted upon, may change in a regular manner. A dependent variable changes as it is acted on by an independent variable outside the system. For example, a change in river discharge (p. 98) (independent variable) causes a change in velocity of flow (dependent variable). *See also* variable[2] (p. 225).

store[1] (*n*) a part of a system (p. 217) that can hold energy or matter, e.g. energy from the Sun held by clouds, and water held in the soil.

loop (*n*) a closed subsystem (↑) showing action on a variable (↑) which then helps that action. There are two sorts of loop: positive feedback (p. 220) loops which are self-destroying, and negative feedback loops which make for equilibrium (p. 220).

element[2] (*n*) a part of something, e.g. in a system (p. 217) an element may be the smallest part possible, or a subsystem (↑). A trade union may be a simple element in a social system, but at a lower level consists of members.

link (*n*) a relation between variables (↑) in a system (p. 217), e.g. a relation is often found between valley side slopes and the slope of the nearby river bed. The nature of the link is often difficult to determine. **link** (*v*), **linkage** (*n*).

regulator (*n*) a key element that controls the speed at which energy passes through a system (p. 217), and which may often be changed by man's action. Infiltration capacity (p. 96) is an example of a regulator within a drainage basin (p. 26) and its effect may be changed by farming practice. **regulate** (*v*).

valve (*n*) a main variable (↑) within a cascading system (p. 217). It may be acted on by man to bring about important changes in the nature of a system (p. 217), e.g. man may change the movement of sand on a beach (p. 46) and bring about great changes in beach shape.

loop

positive feedback loop in a drainage basin

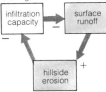

regulator

the effect of a regulator (infiltration capacity) upon the spacing between streams (**D**) and upon valley side slope (**S**)

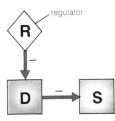

input (*n*) the entry of matter or energy into a
system (p. 217). Examples are: the entry of
hillslope material into a river system, and the
entry of energy into a cascading system
(p. 217).

output (*n*) the mass, energy or change of state
that is produced when an input (↑) passes
through a system (p. 217). An example of
output is the eroded (p. 20) material and
water that moves down and away from a
hillslope.

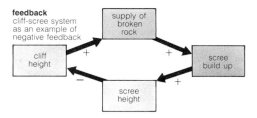

feedback
cliff-scree system
as an example of
negative feedback

feedback (*n*) the way that a system (p. 217) acts
when an input (↑) changes. There are two
kinds of feedback: *negative feedback*, when
the system acts by lessening the effect of the
change, and *positive feedback*, when the
effect of the change is increased. An
ecosystem (p. 72) may act as a negative
feedback system when it preserves its nature
in spite of a change of input, while a glacier
(p. 28) shows positive feedback when its
ability to erode (p. 20) increases when more
snow falls.

equilibrium (*n*) the state of a system (p. 217)
when its input (↑) is equal to its output (↑). As
a result the arrangement of the parts of the
system does not change over time. An
example is a river channel (p. 23) which may
not change over a short time. **equilibria** (*pl*).
See also dynamic equilibrium (↓).

steady state an equilibrium (↑) between the
input (↑) and output (↑) of a system (p. 217).
An important result is that the nature of the
system does not change over time.

dynamic equilibrium
beach (in S. Lincolnshire). Its shape changes over time, but only between the upper and lower profiles

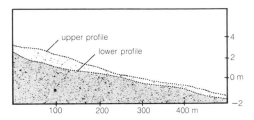

dynamic equilibrium a state of changing equilibrium (↑) between the forces acting on a landform and the material of which it is made. An example is a beach (p. 46), acted on by waves, and whose form differs between summer and winter months, although over many years it shows little change.

damping (*n*) the smoothing effect that takes place when a change in an outside variable (p. 219) is lessened inside a system (p. 217), often by negative feedback (↑).

threshold (*n*) that point at which a steady change in a variable (p. 219) gives rise to an effect that is very different in either nature or degree to that which happened before, e.g. landslides (p. 21) happen in the clay around London when hill slopes increase to a threshold angle of 10°. In central place theory (p. 157), the threshold level of demand is reached when it is worth supplying a particular good (p. 122) or service. This is difficult to measure and often population (p. 146) is used to give demand size. However, similar population sizes may have different demands if their wealth differs.

relaxation (*n*) the path followed by a system (p. 217) from one state of equilibrium (↑) to another. The time taken for the change is known as the relaxation time. This may be quite short, e.g. when a soil responds to a change in plant cover; or long, e.g. when a landscape (p. 22) responds to the change when a former glacier (p. 28) melts. **relax** (*v*).

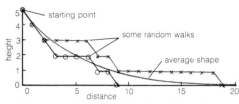

random walk
some random walks used to
make an average down-
valley stream shape

minimum variance the behaviour of a system
(p. 217) that shows the smallest possible
change over time, e.g. a river meander (p. 25)
has that shape which shows least change in
energy output (p. 220) in turning.

minimum variance
shown by a river meander
(the straight line **A** – **B**) and
is an example of the least
work principle (the angle **a** is
very small)

random walk a path made up of steps whose
directions are determined by chance events
that take place in a frame of natural laws, e.g.
river meanders (p. 25) and river networks (p. 26).

stochastic (*adj*) of a system (p. 217) which
works mainly by chance; that is, the laws of
cause and effect do not explain how it
behaves. The spread of new ideas among a
group of people may be stochastic in nature.

equifinality (*n*) the idea that a range of causes
can give rise to similar results, e.g. an erosion
surface (p. 22) whose gentle slopes may result
from any one of a number of causes.

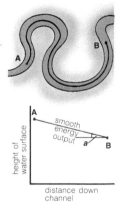

theory (*n*) (1) the ideas that underlie an area of
human activity, especially science; (2) an
explanation put forward to account for the
facts that have been collected. **theorize** (*v*),
theoretical (*adj*).

hypothesis (*n*) a type of theory (↑) which,
although it rests on few facts, provides a
reasonable explanation of a problem.
hypotheses (*pl*), **hypothesize** (*v*),
hypothetical (*adj*).

energy (*n*) the ability to do work. As used, it often
includes various classes of work. **energetic**
(*adj*).

entropy (*n*) a measure of the way energy is
spread out inside a system (p. 217). Entropy
is greatest when energy is spread regularly
through the system, e.g. the peneplain (p. 22)
whose slopes, and so energy, hardly change
over its surface.

model
of a river

sand

indeterminacy
law of flow in a river channel.
Q may be met by many
arrangements of **w, d** and **v**

$$Q = wdv$$

Q = discharge
w = channel width
d = water depth
v = velocity of flow

least work principle the idea that a natural system (p. 217) acts in such a way that the work done is the smallest possible. For example, a river meander (p. 25) is that shape which gives the smallest possible loss of energy (↑) by the flowing water.

inertia (*n*) the tendency of an object to stay at rest or to continue in a certain direction until acted upon by an outside force. It may be a quality of a landscape (p. 22) or human activity that shows little change. **inertial** (*adj*).

dynamic (*adj*) (1) of forces that are not in a state of balance, or (2) of movement that is a result of force.

model (*n*) a simple copy of reality which is used to help our understanding of natural events. There are many types of model. Some may be actually made, (e.g. a flume (p. 99)), and others may be worked out using calculations. Models may be of use in describing the present or future behaviour of systems, for example models of economic (p. 122) development (p. 175). **model** (*v*).

paradigm (*n*) a kind of general model (↑) that lays down the rules that are believed to govern a subject area such as geography (and its various parts), e.g. the cycle of erosion (p. 20) idea provided a paradigm for geomorphology (p. 19) between about 1900 and 1950.

simulation (*n*) a way of copying how a system (p. 217) behaves so that the chief parts of its nature may be studied. This is often done by a computer which may be set up to copy the growth of natural forms such as coastal spits (p. 47) and river networks (p. 26). **simulate** (*v*).

indeterminacy (*n*) the idea that the demands of scientific laws may be met by more than one arrangement of variables (p. 219). *See* diagram opposite. **indeterminate** (*adj*).

Markov chain a number of system (p. 217) states that are related to each other. The change from one state to the next happens in part by chance and in part through the effect of former states. A set of daily weather (p. 67) conditions may form a Markov chain.

statistical analysis the collection, organization and study of numerical data.

analysis (*n*) the process of examining a problem or substance to find out what it consists of, usually to help provide answers to questions.

function (*n*) (1) the normal action of an object; (2) the relationship of one value to another, so that a change in one produces a change in the other.

ratio (*n*) the relation in numbers of one value to another of the same kind.

index (*n*) a measure or pointer of a value. **indices** (*pl*).

graph (*n*) a drawing showing the relationship of one variable (↓) to another. One variable is shown on the upright or x-axis and the other on the y-axis at right angles to it.

histogram (*n*) a diagram that shows the frequency distribution (↓) of data (↓). The diagram is drawn as a multiple-bar graph (↑).

parameter (*n*) a characteristic of a statistical population (p. 146); e.g. measures of central tendency (p. 147), measures of dispersion (p. 84), etc.

data (*n.pl.*) facts which are known, given or obtained.

nominal data data (↑) in name form, having neither quantitative nor relative values; e.g. male, female; pass, fail.

ordinal data data (↑) where only the relative sizes are known and not the actual values, i.e. the data can be ranked or placed in order, either singly or in classes; e.g. small-, middle- and large-size farms.

interval data data (↑) where the actual size or amount is known; e.g. precipitation (p. 55) and population (p. 146) returns.

statistical population a complete set of things, e.g. all the trees in a wood, or all the people in a socio-economic group (p. 150).

sample (*n*) a part of a statistical population (↑), from which data (↑) about the whole population (p. 146) may be obtained. For this purpose the sample should be characteristic of the statistical population. **sample** (*v*).

histogram

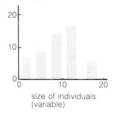

number of individuals (frequency)

size of individuals (variable)

statistical population

spatial distribution of statistical populations

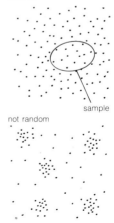

random

sample

not random

discrete variable

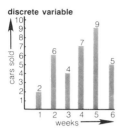

continuous variable

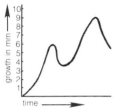

cumulative frequency

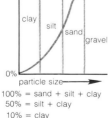

100% = sand + silt + clay
50% = silt + clay
10% = clay

random (*adj*) due to or of chance. **random** (*n*), **randomness** (*n*).

random sample a sample (↑) taken in such a way that each member of the statistical population (↑) or class interval (↓) concerned has an equal chance of being included in the sample.

stratified sample a random sample (↑) of a particular class interval (↓).

quadrat (*n*) the spatial (p. 141) unit into which an area or region (p. 241) is divided for sampling (↑) purposes. It may be of any size or shape, but is often square.

variable² (*n*) a measurable characteristic of any person, place or other thing; e.g. age, size, crop. *See also* variable (p. 219).

discrete variable a variable (↑) which can only take a particular whole number value; e.g. number of cars, number of people. *See also* continuous variable (↓).

continuous variable a variable (↑) which can take any value in a given range (p. 227); e.g. height, depth, distance.

observation (*n*) the value given to each object or thing studied. Thus if 29 measurements of a variable (↑) are made, the observations are numbered 1 to 29.

class interval a grouping of values for a variable (↑) into a class, where the size of the interval is the range (p. 227) of values it includes.

frequency (*n*) the total number of times one value for a variable (↑) appears in all the measurements of that variable. **frequencies** (*pl*).

frequency distribution the frequencies (↑) with which all the values for a variable (↑) appear.

cumulative frequency the frequency (↑) with which a variable (↑) takes values equal to or less than a particular value. It is obtained by adding all the frequencies of lower class intervals (↑).

cumulative (*adj*) of growth or increase by repeated addition.

skew (*n*) a measure of how far values lie more to one side or the other of the mean (↓). When the values lie mainly to the lower skewed end of the range (↓) the condition is described as *positive skewness*. When the opposite happens it is called *negative skewness*. **skew** (*v*), **skewness** (*n*).

kurtosis (*n*) a measure of how flat or peaked a frequency distribution (p. 225) is Where there is a marked peak the distribution is said to be *leptokurtic*, whereas a distribution which is relatively flat is said to be *platykurtic*.

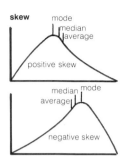

skew

positive skew

negative skew

measures of central tendency

values: 1, 1, 1, 1,   2,   3, 3,   4, 4, 4,   5, 5
frequency   4   1   2   3   2

mean = $\frac{34}{12}$ = 2.83

median = 2.5

mode = 1

measures of central tendency values for a body of data (p. 224) which are characteristic of the data as a whole, e.g. mean (↓), median (↓) and mode (↓) are measures which give the 'centre' of the data.

measures of central location = measures of central tendency (↑).

mean (*n*) the sum of values for all observations (p. 225) divided by the number of observations

average (*n*) = mean (↑).

median (*n*) where observations (p. 225) are ranked (i.e. placed in order) according to value from smallest to largest, the median is the middle observation when the number of observations is odd, and the number half-way between the two middle observations when the number of observations is even.

mode (*n*) the value for a variable (p. 225) which has the highest frequency (p. 225), if there is such a value, or the values of the class containing the highest number of observations, if there is such a class.

measures of dispersion values which relate to the spread, i.e. dispersion, in a set of data (p. 224); e.g. range (↓), interquartile range (↓), standard deviation (↓).

kurtosis

platykurtic

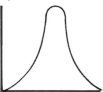

normal peak

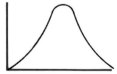

leptokurtic

range (n) the difference between the smallest and largest values in a set of data (p. 224).
interquartile range the difference between the lowest one-quarter and the highest-one-quarter of the values in a set of data (p. 224), i.e. it measures the spread about the median (↑).

interquartile range

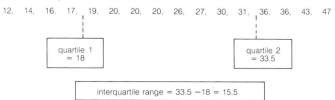

semi-interquartile range half the interquartile range (↑).
deviation (n) the amount by which a value for an observation (p. 225) differs from the mean (↑) of the values for all the observations.
mean deviation the mean (↑) of all the deviations (↑) in a set of data (p 224), regardless of sign, i.e. + or −.
variance (n) the mean (↑) of the squares of the deviations (↑) in a set of data (p. 224).
standard deviation the square root of the variance (↑). This is the most important of the measures of dispersion (↑) for interval data (p. 224).
spatial dispersion the spread of points in geographical space about a centre.
probability (n) the chance of a particular result happening in relation to a particular event,
i.e. $\dfrac{\text{number of times happened}}{\text{number of events}}$
Each of the possible results must be equally likely and independent of each other.
probabilities (pl).
binomial distribution this describes the probabilities (↑) of each of two possible results, which are independent of each other, and which together amount to the sum of all the probabilities, i.e. 100% or 1.0.

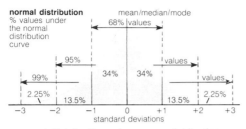

normal distribution
% values under
the normal
distribution
curve

mean/median/mode
68% values
95%
values
99%
34% 34%
values
2.25%
13.5%
13.5%
2.25%

−3 −2 −1 0 +1 +2 +3
standard deviations

normal distribution a frequency distribution
(p. 225) with a bell-shaped curve, slopes
downwards on both sides from the highest
value and 'tails off at both ends. There is no
skew (p. 226) and the mean (p. 226), mode
(p. 226) and median (p. 226) are equal.
About 68% of the area under the curve lies
within ± 1 standard deviation (p. 227) of the
mean, and the probability (p. 227) of a value
being within such a range (p. 227) is therefore
0.68 or 68%. About 95% of the area under the
curve lies within ± 2 standard deviations of
the mean, and the probability of a value being
within such a range is 95% or 0.95.

data transformation the rearrangement of skew
(p. 226) data (p. 224) to fit the normal
distribution for purposes of calculating
probabilities (p. 227), significance testing (↓)
by parametric tests (p. 231), and for some
kinds of correlation (p. 232) and regression
(p. 232), e.g. the data may be transformed into
logarithms, square roots or squares.

sampling error relationship of sampling error to sample size

sampling error

sample size

sampling error the standard deviation (p. 227)
of the sample (p. 224) means (p. 226), used
to calculate the range (p. 227) of the true mean
within given confidence limits (p. 230).

standard error of the sample mean = sampling
error (↑).

data transformation
of skew data

a

original values

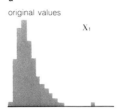

X_1

transformed values

log X_1

b

original values

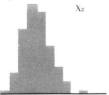

X_2

transformed values

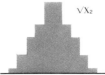

$\sqrt{X_2}$

estimation (*n*) the calculation, as closely as possible from samples (p. 224), of the parameters of a statistical population (p. 146), especially the mean (p. 226) and variance (p. 227), within given confidence limits (p. 230).

determination of sample size the calculation of the number of samples (p. 224) needed to work out the range (p. 227) of the true mean (p. 226) within given confidence limits (p. 230).

determination of sample size
relationship of mean to sample size (the larger the sample the nearer the sample mean to the true mean)

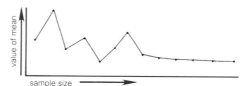

best estimates corrections made to the sampling error (↑) when samples (p. 224) are small.

significance testing calculation of the probability (p. 227) within given confidence limits (p. 230) of a value or unproven idea being true in relation to one or more statistical populations (p. 146), rather than being due to chance arising from the nature of the sample (p. 224) or samples.

hypothesis testing testing to decide: (1) whether differrences between sample (p. 224) means (p. 226), variances (p. 227) and frequencies (p. 225) relate to real differences in the statistical populations (p. 224) concerned; and (2) whether correlations (p. 232) between sample values relate to real correlations in the statistical populations in question. Such testing depends heavily on significance testing (↑).

null hypothesis the unproven idea in hypothesis testing (↑). Usually the null hypothesis is stated in such a way that it is intended not to accept it at a given confidence (p. 230) value, i.e. the hypothesis is rejected (not accepted) if the test proves significant (p. 230).

degrees of freedom the number of independent
observations (p. 225); e.g. if a sample
(p. 224) mean (p. 226) is obtained from 10
observations, the last one is effectively known
once the other 9 have been stated, and so the
degrees of freedom are $10 - 1 = 9$.

confidence (*n*) the amount of trust that can be
placed in a value, unproven idea or a
decision. This is expressed as a probability
(p. 227), e.g. a probability 95% or 0.95 means
that we can have 95% confidence in the value
being correct, or only 5 chances in 100 that it
will be wrong.

significance (*n*) = confidence (↑). **significant** (*adj*).

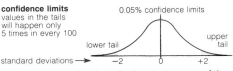

confidence limits
values in the tails
will happen only
5 times in every 100

0.05% confidence limits

lower tail

upper tail

standard deviations →   −2   0   +2

confidence limits quantitative statements of the
confidence that can be placed on a value or
decision. Most often used are the 0.05 and
0.01 confidence limits. The 0.05 confidence
limit means that the value or decision would
be wrong only 5 times in 100. The 0.01
confidence limit means that it would be wrong
only 1 time in 100.

significance levels = confidence limits (↑).

1-tailed test a significance (p. 229) test which is
used where there is good reason to believe
that the sample (p. 224) values are either
smaller or larger than the expected values,
i.e. they are likely to be towards or in just one
'tail' of the frequency distribution (p. 225).
See also 2-tailed test (↓).

2-tailed test a significance (↑) test which is used
when the sample (p. 224) values could be
larger or smaller than the expected values, i.e.
they could be in both 'tails' of the frequency
distribution (p. 225). *See also* 1-tailed test (↑).

comparison tests measures of the amount of
difference or similarity between samples
(p. 224) from two or more statistical
populations (p. 224).

parametric tests comparison tests (↑) for statistical populations (p. 224) which fit the normal distribution, which use observations (p. 225) that are independent of each other, and where the statistical populations under consideration have the same variance (p. 227). Such tests can only be employed when the variables (p. 225) are in the form of interval data (p. 224). Data transformation (p. 228) is sometimes needed. *See also* non-parametric tests (↓).

non-parametric tests comparison tests (↑) which do not depend upon the presence of a particular frequency distribution (p. 225) and which may be used for interval data (p. 224), nominal data (p. 224) and ordinal data (p. 224). Where a normal distribution is known to be skewed (p. 226), non-parametric tests may be employed rather than make a data transformation (p. 228) so that a parametric test (↑) can be performed. Such tests are important in geography because much of the relevant data does not fit the normal distribution.

Student's 't' test a parametric test (↑) to measure the significance (↑) in the difference between two sample (p. 224) mean (p. 226) values, i.e. to help decide whether the difference shows a real difference in the statistical populations (p. 224) concerned or whether they are due to chance arising from the way in which the samples were collected

analysis of variance a parametric test (↑) used to measure the significance (↑) of the differences between three or more sample (p. 224) mean (p. 226) values, i.e. to help decide whether the differences show real differences in the statistical populations concerned or whether they are due to chance arising from the way in which the samples were collected

chi-squared test a non-parametric test (↑) which measures the significance (p. 230) in the difference between actual sample (p. 224) frequencies (p. 225) and those expected according to some null hypothesis (p. 222). The data (p. 224) must be in the form of frequencies.

correlation (*n*) the relationship between two or more variables (p. 225) in time or space, especially the significance (p. 230) of the strength of that relationship. *See also* positive correlation (↓) and negative correlation (↓).

positive correlation a correlation (↑) where an increase in the value of one variable (p. 225) is matched by an increase in the related value of the other.

negative correlation a correlation (↑) where an increase in the value of one variable (p. 225) is matched by a decrease in the related value of the other.

Spearman's rank correlation coefficient a non-parametric (p. 231) measure of correlation (↑).

product moment correlation coefficient a parametric test (p. 231) of correlation (↑).

regression (*n*) the form of the relationship between two variables (p. 225), such that if values for one variable are known, the approximate values for the other may be found. The relationship may take the form of a straight line or a curve.

Poisson distribution a frequency distribution (p. 225) which allows the calculation of the probability (p. 227) of 0, 1, 2, 3, etc. events happening, e.g. the number of quadrats (p. 225) in which 0, 1, 2, 3, etc. examples of a species (p. 81) might be expected. Actual frequencies (p. 225) can thus be related to the Poisson distribution to test for randomness (p. 225) or the lack of it in frequency distributions, including spatial (p. 141) ones.

nearest neighbour analysis an analysis relating the actual mean (p. 226) first-nearest-neighbour distance in a point pattern, e.g. of settlements, to the expected first-nearest-neighbour distance, which is taken to be random (p. 225). It shows whether point patterns have greater or lesser scattering than would be expected by chance.

spatial autocorrelation a systematic (i.e. not random (p. 225)) spatial (p. 141) pattern in measurements or values from a set of points or regions (p. 241) that are next to each other.

correlation
positive correlation
runoff increases
as rainfall increases

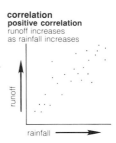

negative correlation
frequency of shops
decreases as distance from
town centre increases

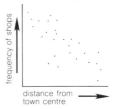

regression
best fit regression line for
observations expressing the
relationship between two
variables **X** and **Y**

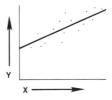

cartography (*n*) the science and skill of map and chart (↓) production including topographic (p. 22), geological and soil maps. Cartographers (↓) use existing maps, triangulation surveys (p. 236) and aerial photographs (p. 205).

cartographer (*n*) a trained and skilled person or scientist who produces maps and charts (↓).

map (*n*) an artistic or graphic representation of features and objects on the Earth's surface. The features may be shown on a flat sheet or on a relief (p. 23) model (p. 223).

chart (*n*) a kind of map used by sailors or pilots. Hydrographic charts show water depth and currents; aeronautic charts show topography (p. 22), hazards (dangers) and flight corridors (paths).

graphicacy (*n*) a name that describes the use of maps, diagrams (figures drawn with lines), photographs and charts (↑) in recording the spatial (p. 141) relationships between objects.

topology (*n*) (1) the scientific study of a locality in detail to determine the relative positions of objects; (2) a form of geometry; (3) a useful method of drawing geographical distribution maps and other maps. **topological** (*adj*).

map projection a way of drawing lines of latitude (p. 238) and longitude (p. 238) on a sheet of paper. Their shapes will be different from those on the globe (p. 238) as it is not possible to copy their global shapes exactly. This means that various sorts of map projection are possible, each with its own characteristics and its own aim at meeting certain needs.

map projection

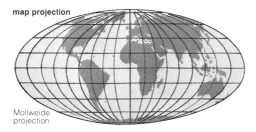

Mollweide projection

Mollweide projection a map projection (p. 233)
named after K. B. Mollweide. It is called a
pseudo-cylindrical projection and the world
appears within an ellipse. The long axis – the
equator (p. 238) is twice the length of the short
axis, the central or standard meridian (p. 238).

Mercator projection

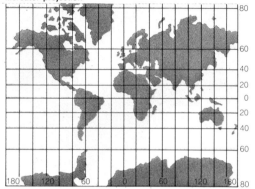

Mercator projection a map projection (p. 233)
named after G. Mercator who first used it in
1569. The parallels (p. 238) and meridians
(p. 238) are all shown as straight lines
crossing at right angles. This projection is used
in setting courses at sea.

Peters' projection a map projection (p. 233)
named after A. D. Peters. The normal parallels
(p. 238) and meridians (p. 238) are replaced
with a grid (p. 236) of 100 decimal degrees.
The projection gives the correct importance to
the densely populated (p. 146) countries.

azimuthal equal-area projection a map
projection (p. 233) in which the spacing of
the parallels (p. 238) decreases the further
they are from the centre of the projection; this
produces an equal-area effect.

map reference a reading of the numbered grid
(p. 236) on a map which is used to define a
position. The reference consists of easting (↓)
and northing (↓) numbers.

grid reference = map reference (↑).

map reference

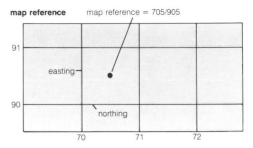

map reference = 705/905

easting

northing

easting (*n*) a measurement eastwards from a point of origin on a map. This is the first part of a map reference (↑).

northing (*n*) a measurement taken northwards of a point of origin on a map (p. 233). This is the second part of a map reference (↑).

map co-ordinates lines that form a grid (p. 236) on a map from which it is possible to define the position of a point on that map.

survey (*n*) the measurement and recording of data (p. 224) of the relief (p. 23) and topography (p. 22) of an area of the Earth's surface. **survey** (*v*), **surveyor** (*n*).

Ordnance Survey a United Kingdom government body, set up in 1791, for the survey, production and publication of topographic (p. 22) maps.

baseline (*n*) (1) a line on the Earth's surface established by very careful measurement; (2) the primary (first) line for the start of a triangulation survey (p. 236).

isohyet (*n*) a line drawn on a map that joins places of equal rainfall.

contour

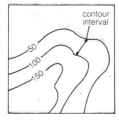

contour interval

isopleth (*n*) a line drawn on a map to join points of equal value, such as rock thickness (isopachyte (p. 236)) or rainfall (isohyet (↑)).

contour (*n*) a line that joins two or more points of the same depth or height on a map or chart (p. 233). *See also* form line (↓).

contour interval the difference in value between successive contours (↓) on a map.

form line a line, usually dotted, drawn where there is too little data (p. 224) to draw a contour (↑).

hachures (*n.pl.*) short lines on a map that show the direction of the maximum slope or a cliff.

grid *see* metric grid (↓).

metric grid a number of lines set at right angles at metric intervals. The grid lines are numbered east and north of a point of origin. The lines cross to form squares and each square can be recognized by its easting (p. 235) and northing (p. 235) numbers.

base net the first figure in a triangulation survey (↓) where the third point is sighted from the two ends of the baseline (p. 235).

triangulation survey a survey (p. 235) based on the measurement of an area using a framework of triangles. The first triangle is taken from a baseline (p. 235) and further angles are measured by use of a theodolite. Such surveys are used for the production of all topographic (p. 22) maps.

triangulation station a sign on a map, consisting of a dot inside a triangle that shows the site of a triangulation survey (↑).

spot height a point on a map with the height printed next to it. The spot usually represents the height above sea-level.

bench mark a permanent height mark used by a surveyor.

isopachyte (*n*) a line used by geologists to join points on a map where a particular rock stratum (p. 18) has the same thickness.

map scale the ratio of the distance measured on a map, to the distance measured between two points on the Earth's surface. A scale of 1 cm to 1 km may be expressed as 1:100,000.

bearing (*n*) a direction of a line taken in relation to the main points of a compass.

cartometric test a way of making sure that the particulars shown on a map or chart (p. 233) are correct.

national grid the metric grid (↑) used by the Ordnance Survey (p. 235).

chart datum level the height of a plane (i.e. level) that is taken as a standard from which all the soundings on a hydrographic chart (p. 98) are taken.

hachures

hachures

metric grid

1 cm to 1 km
scale 1:100,000

metric grid

triangulation station

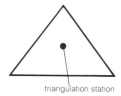

triangulation station

abney level an instrument used to measure slope-angle. It consists of a sighting tube and a spirit level, the bubble of which is seen within the eye piece. When the bubble is exactly over the object sighted, the slope-angle is read from a scale.

altimeter (*n*) an instrument that uses the decrease in atmospheric pressure to show height above sea-level.

planimeter (*n*) (1) a wheeled instrument used in the measurement of distances on maps; (2) an instrument with a movable arm used for measuring areas on maps.

plane table an instrument used by a surveyor to draw maps of a small area. The instrument consists of a tripod (i.e. a three-legged stand) and a flat drawing board. It is used together with an alidade (↓).

alidade (*n*) a simple sight-rule used along with a plane table (↑) during a survey (p. 235).

stereoscopy (*n*) a form of image (p. 203) processing which produces an image that appears solid to the viewer. This is obtained by viewing two images of the same object from different angles with the aid of a stereoscope (↓). **stereoscopic** (*adj*).

stereopairs (*n.pl.*) two overlapping images (p. 203) of the same object or scene which, viewed from different angles, produce an image that appears to be solid. The images are usually black and white photographs viewed through a stereoscope (↓).

stereoscope

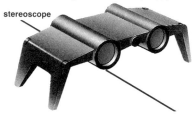

stereoscope (*n*) an instrument used to view two images of the same object from different angles. **stereoscopic** (*adj*).

globe (*n*) (1) the Earth as a whole; (2) a small model (p. 223) of the Earth. **global** (*adj*).

terrestrial (*adj*) (1) of the Earth as a whole. (2) of that part of the Earth above sea-level.

continent (*n*) a very large land area. There are seven continents: Asia (the largest), Africa, North America, South America, Antarctica, Europe and Australia (the smallest). **continental** (*adj*).

hemisphere (*n*) one half of the Earth when it is divided either along the equator (↓) or along a meridian (↓). It usually means the surface area rather than the solid Earth, as in the expression 'southern hemisphere'. **hemispherical** (*adj*).

latitude (*n*) the angular distance, measured in degrees, north or south of the equator (↓). For any place it is the angle between two lines drawn from that place and from the equator to the centre of the Earth.

equator (*n*) an imaginary line around the Earth which is an equal distance from the two poles (↓). It is the longest line of latitude (↑), and the only one that is a great circle (↓). **equatorial** (*adj*).

parallel (*n*) (1) an imaginary line on the Earth's surface which runs at the same distance from the equator (↑); (2) any line which runs at the same distance from another.

longitude (*n*) the angular distance measured in degrees west or east of a prime meridian (↓). All lines of longitude pass through the poles (↓).

meridian (*n*) an imaginary line that runs from pole (↓) to pole on the surface of the Earth.

prime meridian a meridian (↑) that is used as a starting line for measuring the longitude (↑) of any point on the Earth's surface. The one most often used is the Greenwich Meridian (↓).

Greenwich Meridian the meridian (↑) that passes through Greenwich, England, and from which the angular distance in degrees of any point on the Earth's surface may be measured. It has a longitude (↑) of 0° and is the most often used prime meridian (↑).

latitude

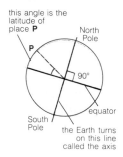

this angle is the latitude of place **P**

the Earth turns on this line called the axis

equator

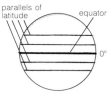

parallels of latitude

longitude

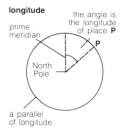

the angle is the longitude of place **P**

prime meridian

a parallel of longitude

meridian

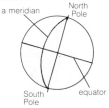

a meridian

north
the relation between grid
north and true north

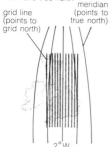

Tropics of Cancer and Capricorn

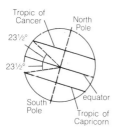

pole (*n*) one of two points where the axis (i.e. the line on which the Earth turns) cuts the surface of the globe. These two points are called the North and South Poles. Poleward means 'towards the poles'. **polar** (*adj*).

north (*n*) the direction of the North Pole from any point on the Earth's surface. This direction is also called *true north*. *Magnetic north* is the direction in which a compass needle points. From Britain, it is about 6° west of true north. *Grid north* is the northerly direction of the lines of the British National Grid. Each of these lines runs parallel to the meridian (↑) 2° west of the Greenwich meridian (↑), and so makes an angle with all other meridians.

great circle a circle that is the same as the circumference of the Earth. The shortest distance between two places on the surface of the Earth lies on the great circle that connects them.

Tropic of Cancer a line of latitude (↑) that lies 23½° north of the equator (↑) and where the Sun is directly overhead at midsummer in the northern hemisphere (↑).

Tropic of Capricorn a line of latitude (↑) that lies 23½° south of the equator (↑) and where the sun is directly overhead at midsummer in the southern hemisphere (↑).

Arctic circle a line of latitude (↑) drawn at 66½° north. Between this line and the North Pole, is the Arctic region where night lasts for 24 hours in the middle of winter. In the middle of summer, daylight lasts for 24 hours (the 'midnight sun').

Antarctic circle a line of latitude (↑) drawn at 66½° south. Between this line and the South Pole the conditions are similar to those between the Arctic circle (↑) and the North Pole.

Arctic and Antarctic circles

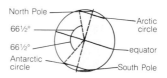

International Date Line

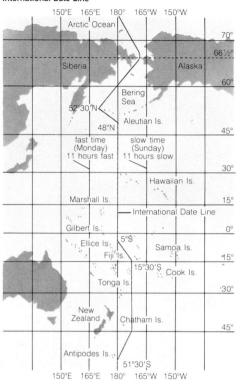

International Date Line an imaginary line on the
Earth's surface which, when crossed, brings
about a change of exactly one day in the date.
Movement across the line from east to west
causes the date to change suddenly to that of
a day later; i.e. a day is lost. Movement from
west to east across the line causes a day to
be gained. This line is necessary because of
the way that time changes east (earlier) and
west (later) of the Greenwich Meridian (p. 238).
The line, which passes through the Pacific
Ocean, avoids populated (p. 146) islands.

zone (*n*) (1) a large part of the Earth's surface that has a similar character everywhere. This may be climatic (p. 71) or vegetational (p. 86). It usually forms a band running west to east around the Earth. (2) A smaller area of the Earth's surface that has a similar character everywhere, e.g. industrial zone (p. 167), transition zone (p. 168). (3) A part of the Earth below the surface that has a similar character everywhere, e.g. Benioff zone (p. 16). **zonal** (*adj*).

polar zone one of the two zones (↑) that lie north of the Arctic circle (p. 239) or south of the Antarctic circle (p. 239). The polar zone is characterized by the presence of permafrost (p. 36) and of pack ice (large areas of ice covering all the surface of the sea).

temperate zone that part of the Earth's surface which experiences temperature and rainfall conditions of a kind between those of the polar zones (↑) and tropical (↓) zones (↑).

tropics (*n*) (1) that part of the Earth's surface that lies between latitudes (p. 238) 23½° north and 23½° south of the equator (p. 238); that is, the zone (↑) where the Sun is overhead for part of the year; (2) a zone characterized by a certain climate (p. 71). For example, the yearly isotherm (p. 54) of 20°C has been used as the northern and southern edge of the tropics. The sub-tropics are regions (↓) that lie just north of the Tropic of Cancer (p. 239) and south of the Tropic of Capricorn (p. 239). **tropical** (*adj*).

equatorial zone a biome (p. 75) that covers about 8% of the Earth's land surface on and near the equator (p. 238). It has high temperatures and rainfall, and is largely made up of rainforest (p. 87).

region (*n*) a part of the Earth's surface, one of whose characteristics remains the same everywhere. This characteristic may be related to physical geography (p. 8), e.g. climatic (p. 71), or it may be related to human geography (p. 106), e.g. the use made of the land. **regional** (*adj*).

Some useful abbreviations and acronymns

ADB	Asian Development Bank
AONB	Area of Outstanding Natural Beauty (UK)
ASEAN	Association of South East Asian Nations
BART	Bay Area Rapid Transit (system) (San Francisco, USA)
BSC	British Steel Corporation
CACEU	Central African Customs and Economic Union
CACM	Central American Common Market
CAP	Common Agricultural Policy
CBD	central business disttrict
COMECON	Council for Mutual Economic Aid
CoSIRA	Council for Small Industries in Rural Areas (UK)
DTI	Department of Trade and Industry (UK)
EAC	East African Community (Kenya, Uganda, Tanzania)
ECA	Economic Commission for Africa
ECAFE	Economic Commission for Asia and the Far East
ECAP	Economic Commission for Asia and the Pacific
ECLA	Economic Commission for Latin America
ECOWAS	Economic Community of West African States
EDA	Economic Development Agency
EDR	economic development region
EE	English Estates
EEC	European Economic Community
EFTA	European Free Trade Area
EIS	Environmental Impact Statement (USA)
EPA	Environmental Protection Agency (USA)
ERTS	Earth Resources Technology Satellite
fob	free on board
FTZ	foreign trade zone
GATT	General Agreement on Tariffs and Trade
GDP	gross domestic product
GNP	gross national product
GOES	Geostationary Operational Environmental Satellite
HDC	highly developed country
HST	high speed train

IBRD	International Bank for Reconstruction and Development
ITCZ	intertropical convergence zone
LAFTA	Latin American Free Trade Association
LDC	less developed country
LLDC	least developed country
LNG	liquified natural gas
MDC	moderately developed country
MNC	multinational corporation or company
NAWAPA	North American Water and Power Alliance
NEPA	National Environmental Policy Act (1969) (USA)
NIC	newly industrializing country
NIEO	New International Economic Order
NPA	National Park Authority (UK)
OECD	Organization for Economic Co-operation and Development
OPEC	Organization of Petroleum Exporting Countries
SDA	Scottish Development Agency
SMLA	standard metropolitan labour area (UK)
SMSA	standard metropolitan statistical area (USA)
SST	supersonic transport plane
TNC	transnational company
TPC	territorial production complex
TVA	Tennessee Valley Authority (USA)
UNCLOS	United Nations Conference on the Law of the Sea
UNCTAD	United Nations Conference on Trade and Development
UNDP	United Nations Development Programme
UNESCO	United Nations Educational and Scientific Organization
UNIDO	United Nations Industrial Development Organization
VLCC	very large crude carrier
WDA	Welsh Development Agency
WHO	World Health Organization
ZEG	zero economic growth
ZPG	zero population growth

Geological time – major divisions

era and sub-era	period		epoch	ages (Ma) millions of years
Cenozoic	Quaternary		Holocene	0.01
			Pleistocene	2.0
	Tertiary	Neogene	Pliocene	5.1
			Miocene	24.6
		Palaeogene	Oligocene	38.0
			Eocene	54.9
			Palaeocene	65.0
Mesozoic	Cretaceous			144
	Jurassic			213
	Triassic			248
Palaeozoic	Permian			286
	Carboniferous			360
	Devonian			408
	Silurian			438
	Ordovician			505
	Cambrian			590
	Pre-Cambrian			

International System of Units (SI)

PREFIXES

PREFIX	FACTOR	SIGN	PREFIX	FACTOR	SIGN
milli	$\times 10^{-3}$	m	kilo	$\times 10^{3}$	k
micro	$\times 10^{-6}$	μ	mega	$\times 10^{6}$	M
nano	$\times 10^{-9}$	n	giga	$\times 10^{9}$	G
pico	$\times 10^{-12}$	p	tera	$\times 10^{12}$	T

BASE UNITS

UNIT	SYMBOL	MEASUREMENT
metre	m	length
kilogram	kg	mass
second	s	time
ampere	A	electric current
kelvin	K	temperature
mole	mol	amount of substance
candela	cd	luminous intensity

COMMON DERIVED UNITS

UNIT	SYMBOL	MEASUREMENT
newton	N	force
joule	J	energy, work
hertz	Hz	frequency
pascal	Pa	pressure
coulomb	C	quantity of electric charge
volt	V	electrical potential
ohm	Ω	electrical resistance

Index

Physical Map of the World

Hammer Equal Area projection